Advances in Transmission Electron Microscopy for the Study of Soft and Hard Matter

Advances in Transmission Electron Microscopy for the Study of Soft and Hard Matter

Editor

Elvio Carlino

MDPI • Basel • Beijing • Wuhan • Barcelona • Belgrade • Manchester • Tokyo • Cluj • Tianjin

Editor
Elvio Carlino
Institute for Microelectronics and Microsystems (IMM)
Italy

Editorial Office
MDPI
St. Alban-Anlage 66
4052 Basel, Switzerland

This is a reprint of articles from the Special Issue published online in the open access journal *Materials* (ISSN 1996-1944) (available at: https://www.mdpi.com/journal/materials/special_issues/advances_TEM).

For citation purposes, cite each article independently as indicated on the article page online and as indicated below:

LastName, A.A.; LastName, B.B.; LastName, C.C. Article Title. *Journal Name* **Year**, *Volume Number*, Page Range.

ISBN 978-3-0365-3212-7 (Hbk)
ISBN 978-3-0365-3213-4 (PDF)

Cover image courtesy of Elvio Carlino.

© 2022 by the authors. Articles in this book are Open Access and distributed under the Creative Commons Attribution (CC BY) license, which allows users to download, copy and build upon published articles, as long as the author and publisher are properly credited, which ensures maximum dissemination and a wider impact of our publications.
The book as a whole is distributed by MDPI under the terms and conditions of the Creative Commons license CC BY-NC-ND.

Contents

About the Editor vii

Preface to "Advances in Transmission Electron Microscopy for the Study of Soft and Hard Matter" ix

Elvio Carlino
Special Issue: Advances in Transmission Electron Microscopy for the Study of Soft and Hard Matter
Reprinted from: *Materials* **2021**, *14*, 1711, doi:10.3390/ma14071711 1

Sylvain Trépout
Tomographic Collection of Block-Based Sparse STEM Images: Practical Implementation and Impact on the Quality of the 3D Reconstructed Volume
Reprinted from: *Materials* **2019**, *12*, 2281, doi:10.3390/ma12142281 5

Elvio Carlino
In-Line Holography in Transmission Electron Microscopy for the Atomic Resolution Imaging of Single Particle of Radiation-Sensitive Matter
Reprinted from: *Materials* **2020**, *13*, 1413, doi:10.3390/ma13061413 19

Tatiana Latychevskaia
Holography and Coherent Diffraction Imaging with Low-(30–250 eV) and High-(80–300 keV) Energy Electrons: History, Principles, and Recent Trends
Reprinted from: *Materials* **2020**, *13*, 3089, doi:10.3390/ma13143089 39

Lluís López-Conesa, Carlos Martínez-Boubeta, David Serantes, Sonia Estradé and Francesca Peiró
Mapping the Magnetic Coupling of Self-Assembled Fe_3O_4 Nanocubes by Electron Holography
Reprinted from: *Materials* **2021**, *14*, 774, doi:10.3390/ma14040774 75

Lahcen Khouchaf, Khalid Boulahya, Partha Pratim Das, Stavros Nicolopoulos, Viktória Kovács Kis and János L. Lábár
Study of the Microstructure of Amorphous Silica Nanostructures Using High-Resolution Electron Microscopy, Electron Energy Loss Spectroscopy, X-ray Powder Diffraction, and Electron Pair Distribution Function
Reprinted from: *Materials* **2020**, *13*, 4393, doi:10.3390/ma13194393 85

Minkyo Jung, Hyosun Choi, Jaekwang Kim and Ji Young Mun
Correlative Light and Transmission Electron Microscopy Showed Details of Mitophagy by Mitochondria Quality Control in Propionic Acid Treated SH-SY5Y Cell
Reprinted from: *Materials* **2020**, *13*, 4336, doi:10.3390/ma13194336 99

Stefan Löffler, Michael Stöger-Pollach, Andreas Steiger-Thirsfeld, Walid Hetaba and Peter Schattschneider
Exploiting the Acceleration Voltage Dependence of EMCD
Reprinted from: *Materials* **2021**, *14*, 1314, doi:10.3390/ma14051314 109

About the Editor

Elvio Carlino is an Italian physicist, and he currently serves as Research Director at the Italian National Research Council (CNR). Since 1985, his main interest has been in the field of electron microscopy for the study of matter. His research activity has been dedicated to the application and developments of new transmission electron microscopy methodologies for studies of the structural, chemical, electronic, and magnetic properties of solids.

Preface to "Advances in Transmission Electron Microscopy for the Study of Soft and Hard Matter"

This book is focused on advances in transmission electron microscopy (TEM) for the study of matter. These advances are certainly, but not only, due to the technological improvement of the equipment. The advances are also due to the development of original approaches in the use of the equipment and of the results of relevant experiments. These new approaches can enable new ways to study the properties of matter. In the rather long history of electron microscopy, such advances have taken place many times, widening our view of Nature's secrets. The aim here is to provide a few recent hints, not only to the specialists on TEM, such as those who contributed to this volume, but also to the wide audience of those who can extend their knowledge of the properties of their specimens via transmission electron microscopy experiments. Finally, let me acknowledge the contributors to this Special Issue. I would also like to thank Ms. Charlene Liao for her assistance with the editorial work. A special thanks to Dr. Antonietta Taurino for the many contributions during the preparation of this work.

Elvio Carlino
Editor

Editorial

Special Issue: Advances in Transmission Electron Microscopy for the Study of Soft and Hard Matter

Elvio Carlino

Istituto per la Microelettronica ed i Microsistemi, Consiglio Nazionale delle Ricerche (CNR-IMM), Sezione di Lecce, Campus Universitario, via per Monteroni, 73100 Lecce, Italy; elvio.carlino@cnr.it

Citation: Carlino, E. Special Issue: Advances in Transmission Electron Microscopy for the Study of Soft and Hard Matter. *Materials* **2021**, *14*, 1711. https://doi.org/10.3390/ma14071711

Received: 22 March 2021
Accepted: 24 March 2021
Published: 31 March 2021

Publisher's Note: MDPI stays neutral with regard to jurisdictional claims in published maps and institutional affiliations.

Copyright: © 2021 by the author. Licensee MDPI, Basel, Switzerland. This article is an open access article distributed under the terms and conditions of the Creative Commons Attribution (CC BY) license (https://creativecommons.org/licenses/by/4.0/).

Transmission Electron Microscopy (TEM) owes its success to the capability to investigate fundamental aspects of nature, answering the human need of knowledge necessary to understand unknown mechanisms and to find new solutions in a variety of fields like physics, biology, medicine, engineering, or chemistry. Since the beginning of modern science, the scientist necessitated to see, in a general sense, the details of a phenomenon to imagine and to develop a model capable of explaining the phenomenon itself. From this point of view, a microscope is the archetype tool capable of studying the ultimate elements of phenomena, which are invisible to the naked eye. When the scientific interest is focused on an atomic scale, this archetypic tool finds its highest expression in the transmission electron microscope. It is worthwhile to remark that the electron microscope itself is nearly useless alone, as it needs microscopy, which is the powerful combination of the most advanced technological equipment for imaging, diffraction, and spectroscopies with the knowledge and the methods necessary to explore all the opportunities provided by the microscope and by the depth of the strong electron-matter interaction. In fact, it is electron microscopy that provides answers to fundamental physical questions, such as the experimental demonstration of the self-interference of the electron [1,2], previously believed possible only as a *gedanken* experiment proposed by Albert Einstein (Richard Feynman is said to have re-marked that self-interference of the electron is the phenomenon that contains everything you need to know about quantum mechanics). Again, it is electron microscopy that provides a vast variety of applications to the study of organic and inorganic samples at an atomic resolution, investigating their shape, crystal structure, chemistry, electric properties, and magnetic properties. From the long story of electron microscopy, it emerges that the conventional use of an electron microscope is a fruitful way to investigate the matter by well-established powerful methodologies, whereas an unconventional use of an electron microscope could sometimes open new routes to new, unexpected knowledge.

This special issue was conceived with this idea in mind, focusing on the advances in transmission electron microscopy, and also scanning transmission electron microscopy (STEM) for the study of both organic and inorganic matter. The purpose is to offer to the scientific community an opportunity to show some of the latest developments in TEM/STEM based methodologies. However, we are conscious that a single issue can cover only a few aspects of this field of investigation. Here, a particular attention is paid to the radiation damage in TEM/STEM experiments, which is currently one of the most limiting factors to the further improvement in spatial resolution and accuracy in atomic resolution imaging and spectroscopy experiments, not only on biological samples.

Within works on approaches to reduce the dose delivered to a specimen, there is a paper [3] dedicated to electron tomography in STEM. Here, electron tomography is applied to seize the global overview of cellular architecture in 3D at the nanometer scale. The need to collect several images with meaningful contrast for 3D tomography is evident in competition with the radiation damage of the sample. Since sparse data collection can perform efficient electron dose reduction, whereas the risk is to lose some information, in Reference [3], the author proposes a method based on compressing sensing or inpainting algorithms for the missing information reconstruction. The method is, hence, applied, as a case study, to a thick biologic specimen.

The paper [4] is an example of how an unconventional use of a standard TEM, equipped with a high coherent electron source, can provide a way to overcome the radiation damage during atomic resolution experiments on single nanoparticles of radiation sensitive, organic, or inorganic matter consisting of low atomic number elements. For these specimens, the nanometric size of the particles and the low scattering power of their constituents jeopardize even the overview of the specimens. In Reference [4], it is demonstrated how in-line holography imaging, performed by conveniently tuning the electron optical conditions, can provide a high contrast overview of the specimen, while delivering a low-density current of electrons of a few $e^-\text{Å}^{-2}s^{-1}$. Furthermore, the in-line holograms can be used to tune the electron optical conditions to enable a low dose atomic resolution phase contrast imaging to study the properties of single particles of nano-drugs or of biologic matter.

In-line electron holography, off-axis electron holography, point projection electron microscopy, and electron coherent diffraction imaging are the subjects of paper [5]. In fact, Reference [5] is a review on the theoretical background necessary to understand these approaches and on the recent theoretical and experimental advances in these fields In view of their importance in the study of radiation-sensitive materials, the significant role of the electron energy has been considered and two ranges of energies of applicative relevance (30–250 eV and 80–300 keV) have been exploited in detail, discussing advantages and disadvantages of the choice of a specific energy, as a function of the specimen of interest. Finally, an explicit comparison between electron holography and electron coherent diffraction imaging has been made both for their capabilities to measure the phase of the electron waves scattered by the specimen and in terms of a minimum dose delivered to the samples.

The paper [6] is an example of application of off-axis holography to the study of magnetic properties of nanoparticles, especially to access and map the magnetic configuration of Fe_3O_4 cubic nanoparticles for potential application in magnetic hyperthermia, as a complementary approach to standard therapies for cancer treatment. The advances in the equipment for off-axis holography experiments, providing multiple biprisms for accurate tuning of the field of view and of the experimental setting, enable quantitative mapping of the magnetic properties of single nanoparticles in relationship with the other particles, resulting in the formation of chains, whose shape and size have direct influence for the medical applications. The accuracy of the magnetic mapping makes possible an appropriate comparison with simulations, which is necessary to unveil the complexity of this matter.

One of the reasons for the success of TEM is the possibility to perform several kinds of experiments on the same sample in the same instrument, gaining pieces of cross correlated information, whose ensemble enables us to reach a degree of accuracy and confidence in the knowledge of the properties of a complex specimen, not reachable in a single kind of experiment. The challenging study of the microstructure of amorphous silica was embarked on paper [7] by HRTEM, Electron Energy Loss Spectroscopy and Electron Pair Distribution Function complemented by X-ray powder diffraction.

A single TEM image or spectrum can achieve accurate atomic resolution information on a nanometric volume of the specimen. As a consequence, the experimental strategy of a successful TEM investigation requires us to explore a representative number of regions of interest within the same TEM specimen, and a representative number of TEM specimens, to investigate a general property of the matter under study, and not only a local feature seen by accident on a TEM specimen. Correlative light microscopy and TEM studies are not trivial from an experimental point of view but allow one to complement the peculiarities of two approaches that merge information on the same area achieved with the relevant spatial resolutions. This is what has been investigated in Reference [8], where correlative light microscopy and TEM are successfully applied to the study of selective degradation of mitochondria by autophagy, following the process on a nanometric scale, in cells under stress.

The paper [9] is an example of study on electron energy loss magnetic chiral dichroism (EMCD). EMCD was experimentally demonstrated in 2006 [10], and it is analogous to the X-ray chiral circular dichroism (XMCD), which is an approach, developed about 20 years before, that enabled us to quantitatively study the magnetic phenomena in a correlated electron system by using circularly polarized x-ray photons in a synchrotron. The born of EMCD is another valuable example of how an unconventional use of a TEM can open new ways to understand the nature. The origin of EMCD is related to the observation that the absorption cross section for X-rays and electrons are similar, if we replace the polarization vector for photon absorption with the exchanged momentum in electron impact ionization. This paved the way to discover a method to study the magnetic properties of correlated electrons in solids with a spatial resolution on an atomic scale, which is typical of a TEM. Since the proof of concept of EMCD, the experimental and theoretical advances in this field made EMCD a method currently used all over the world to quantitatively study the magnetic properties of the matter. The paper [9] focuses on the dependence of EMCD on the acceleration voltage and how this basic experimental parameter can be used to optimize EMCD experiments. This is done by deriving an analytic formula for predicting EMCD effects and elucidating the underlying physics, which enables a better tailoring of the electron optical conditions for quantitative EMCD.

Funding: This research received no external funding.

Conflicts of Interest: The authors declare no conflict of interest.

References

1. Crease, R.P. The most beautiful experiment 2002. *Phys. World* **2002**, *15*, 19.
2. Merli, P.G.; Missiroli, G.F.; Pozzi, G. On the statistical aspect of electron interference phenomena. *Am. J. Phys.* **1976**, *44*, 306–307. [CrossRef]
3. Trépout, S. Tomographic Collection of Block-Based Sparse STEM Images: Practical Implementation and Impact on the Quality of the 3D Reconstructed Volume. *Materials* **2019**, *12*, 2281. [CrossRef] [PubMed]
4. Carlino, E. In-Line Holography in Transmission Electron Microscopy for the Atomic Resolution Imaging of Single Particle of Radiation-Sensitive Matter. *Materials* **2020**, *13*, 1413. [CrossRef] [PubMed]
5. Latychevskaia, T. Holography and Coherent Diffraction Imaging with Low-(30–250 eV) and High-(80–300 keV) Energy Electrons: History, Principles, and Recent Trends. *Materials* **2020**, *13*, 3089. [CrossRef] [PubMed]
6. López-Conesa, L.; Martínez-Boubeta, C.; Serantes, D.; Estradé, S.; Peiró, F. Mapping the Magnetic Coupling of Self-Assembled Fe_3O_4 Nanocubes by Electron Holography. *Materials* **2021**, *14*, 774. [CrossRef]
7. Khouchaf, L.; Boulahya, K.; Das, P.P.; Nicolopoulos, S.; Kovács Kis, V.; Lábár, J.L. Study of the Microstructure of Amorphous Silica Nanostructures Using High-Resolution Electron Microscopy, Electron Energy Loss Spectroscopy, X-ray Powder Diffraction, and Electron Pair Distribution Function. *Materials* **2020**, *13*, 4393. [CrossRef]
8. Jung, J.; Choi, H.; Kim, J.; Mun, J.Y. Correlative Light and Transmission Electron Microscopy Showed Details of Mitophagy by Mitochondria Quality Control in Propionic Acid Treated SH-SY5Y Cell. *Materials* **2020**, *13*, 4336. [CrossRef] [PubMed]
9. Löffler, S.; Stöger-Pollach, M.; Steiger-Thirsfeld, A.; Hetaba, W.; Schattschneider, P. Exploiting the Acceleration Voltage Dependence of EMCD. *Materials* **2021**, *14*, 1314. [CrossRef]
10. Schattschneider, P.; Rubino, S.; Hebert, C.; Rusz, J.; Kunes, J.; Novák, P.; Carlino, E.; Fabrizioli, M.; Panaccione, G.; Rossi, G. Detection of magnetic circular dichroism using a transmission electron microscope. *Nature* **2006**, *441*, 486–488. [CrossRef] [PubMed]

Article

Tomographic Collection of Block-Based Sparse STEM Images: Practical Implementation and Impact on the Quality of the 3D Reconstructed Volume

Sylvain Trépout

Institut Curie, Inserm U1196, CNRS UMR 9187, Université Paris Sud, Centre Universitaire, Bât. 110-112, 91405 Orsay CEDEX, France; sylvain.trepout@curie.fr; Tel.: +33-169-86-30-81

Received: 15 June 2019; Accepted: 11 July 2019; Published: 16 July 2019

Abstract: The reduction of the electron dose in electron tomography of biological samples is of high significance to diminish radiation damages. Simulations have shown that sparse data collection can perform efficient electron dose reduction. Frameworks based on compressive-sensing or inpainting algorithms have been proposed to accurately reconstruct missing information in sparse data. The present work proposes a practical implementation to perform tomographic collection of block-based sparse images in scanning transmission electron microscopy. The method has been applied on sections of chemically-fixed and resin-embedded *Trypanosoma brucei* cells. There are 3D reconstructions obtained from various amounts of downsampling, which are compared and eventually the limits of electron dose reduction using this method are explored.

Keywords: scanning transmission electron microscopy (STEM); electron tomography (ET); sparse imaging; inpainting reconstruction; biological samples; *Trypanosoma brucei*

1. Introduction

Accessing the ultrastructure and cellular organization of cell components has always been of great help in deciphering their functions and mechanisms inside the cell [1–4]. Electron tomography (ET) is a powerful tool to seize the global overview of cellular architecture in 3D at the nanometer scale. However, transmission electron microscopy (TEM)-based methods are limited to the study of thin samples because the substantial electron scattering occurring in thick samples leads to the collection of images with weak contrast and poor signal-to-noise ratio (SNR). Even though more detailed and contrasted images can be obtained by filtering out inelastically scattered electrons using energy filters, above about 300 nm (or little more depending on the acceleration voltage of the microscope) TEM-based methods are inefficient to capture quality images. Other methods have been developed to image thick specimens. Focused-ion beam (FIB) milling [5–7] only reduces the thickness of thick samples so that they become thin enough, but do not allow to image thick samples. Serial sectioning aims at imaging thick samples but it is a tedious task [8]. An alternative to TEM, is the use of scanning transmission electron microscopy (STEM) that is based on the raster scanning of an electron beam focused on the sample, the transmitted electrons being collected by one (or more) post-specimen detector(s) [9,10]. Thanks to the beam geometry; through the point-to-point imaging pattern and the absence of electromagnetic lens post-specimen, it has been shown that STEM is able to produce high SNR and high contrast images of thick biological samples [11,12]. As opposed to TEM, STEM is an incoherent imaging mode, less affected by the strong scattering of the electron waves occurring in thick samples. Importantly it has been shown that bright-field imaging produces images of higher fidelity compared to annular dark-field because of the multiple electron scattering that can occur inside the sample [13,14]. It has also been shown on cryo-preserved prokaryotic cells that at equivalent electron doses, the use of STEM generates higher SNR images compared to TEM while reducing structural damages [15]. More recently, simulations have shown that biological samples of micron thicknesses ("and beyond") could be studied in cryo-STEM tomography [16]. However, studying thick specimens involves the use of high electron doses to maintain a sufficient number of electrons at the detector level, increasing radiation damages. Strategies to reduce the electron dose are of main significance to preserve the sample integrity.

In ET, several strategies based on sparse acquisition (i.e., downsampling) of the data have been proposed to reduce radiation damages. Restoration of missing information is performed using data compression approaches based on algorithms such as compressive-sensing (CS) or inpainting. In previous work, several algorithms such as discrete algebraic reconstruction technique [17], total variation minimization [18], CS [19–21] have been developed. In these different works, the downsampling is performed either by (i) reducing the number of acquired tilt angles or by (ii) reducing the number of pixels collected per image (i.e., sparse images). Very recently, the limits of the reduction of collected tilt-angles associated with sparsity-exploiting reconstructions have been explored and show that they are highly dependent on the sample properties [22]. More precisely, specimen complexity and noise severely degrade reconstruction quality in sparsity-exploiting methods. Biological samples are beam sensitive complex systems requiring the use of a low electron dose, hence the collection of noisy images. For these reasons, strategies reducing the number of collected tilt-angles perform very well on samples with low complexity, whereas they do not perform so well on biological samples because of the very intricate state of beam-sensitive biological matter. Whereas, it is trivial to reduce the number of acquired tilt angles in TEM or STEM tomographic experiments, the actual implementation of sparse imaging at the level of the image itself is much more complex. In TEM, there is no strategy yet to collect sparse images whereas STEM has the potential to collect sparse images thanks to its point-to-point imaging pattern as shown for example in a STEM-Electron Energy Loss Spectroscopy experiment [23]. Despite the possibility to collect sparse images in STEM, most of the works studying sparsity-exploiting strategies simulate sparse images in silico post-acquisition [19,21,24,25]. In rare experiments, sparse images were experimentally collected, the electron beam being either blanked using a fast electromagnetic shutter at the level of the condenser [26] or being driven using a dedicated scan system [27,28]. Very recently, Vanrompay et al. used a fast electromagnetic shutter to perform sparse imaging in 3D on gold nanoparticles [29].

There is a growing interest in biology to use sparse acquisitions in ET experiments [30,31] and the need to develop such acquisition methods is urgent. The goal of the present work is to develop a tomographic acquisition method able to collect sparse images to study beam-sensitive thick biological samples in STEM. Section 2 describes the materials used and the electron microscope setup. Section 3.1 presents first the practical implementation of sparse images in STEM on 2D images. Section 3.2 extends the method to perform sparse imaging during a tilt-series to perform 3D imaging. In Section 3.3 the method is applied on a 500 nm-thick chemically-fixed and resin-embedded *Trypanosoma brucei* sample. The pros and cons of sparse imaging using the scanning coils of the electron microscope and the effect of the electron dose reduction on the quality of the 3D reconstruction are discussed thereafter.

2. Materials and Methods

2.1. Test Samples

The validation of the practical implementation to collect 2D sparse images has been performed on a carbon crossed line grating replica grid and on iron oxyde nanoparticles [32]. The carbon replica used in this work had evaporated Au/Pd atoms and latex spheres at the surface (EMS #80055). The development and the validation of the 3D tomography workflow was performed on the carbon crossed line grating replica grid.

After validation of the tomography workflow, it was applied on *T. brucei* resin Sections. *T. brucei* cells (strain 427) were cultured in SDM79 medium supplemented with hemin and 10% fetal calf serum [33]. Cells were chemically fixed directly in the culture medium with 2.5% glutaraldehyde for 30 min, exhaustively rinsed in phosphate-buffered saline, post-fixed in 2% OsO4 for 30 min at 4 °C in the dark, dehydrated in baths of increasing ethanol concentrations, and embedded in Epon as previously described [34]. After resin polymerization at 60 °C for 48 h, resin blocks were cut to produce sections of 500 nm which were deposited on bare 200 mesh copper grids (EMS #G200-Cu).

2.2. Experimental Setup of the Electron Microscope

The 200 kV field emission gun electron microscope (2200FS, JEOL, Tokyo, Japan) in scanning mode was set up to enable imaging conditions for thick specimens using the following parameters (condenser lens aperture: 40 µm, camera length: 80 cm, semi-convergence angle was 9.3 mrad and outer bright-field semi-collection angle was 5 mrad). Images collected between 20,000 × and 30,000 × magnifications (corresponding pixel sizes were 2.08 and 1.38 nm respectively) were digitized using a Digiscan II ADC module (16 bits). The beam current was 2.2 pA.

2.3. Tilt-Series Collection, Alignment and Reconstruction

Tilt-series on the crossed line grating replica grid were collected at 30,000 × magnification, between −65° and +63° using 0.5° tilt increments and the dwell time was set to 8 µs. Using these collection conditions, the final electron dose received by the sample after completion of the tilt-series has been estimated around 1000 e$^-$/Å^2 for fully collected images. These settings have been used to generate high contrast images. During the tilt-series, (i) a sparse image downsampled to 12.5% of the total amount of pixels and (ii) a fully collected image were collected at each tilt angle. Fully collected images were used as ground truth images. Tilt-series on the T. brucei resin section were collected at 20,000 × magnification, between −67° and +68° using 1° tilt increments and the dwell time was set to 1 µs. These settings correspond to a significant electron dose reduction compared to the settings used on the crossed line grating test sample. The electron dose was deliberately diminished to match very low dose collection conditions. Using these collection conditions, the final electron dose received by the sample after completion of the tilt-series has been estimated around 70 e$^-$/Å^2 for fully collected images. To study the impact of the electron dose reduction on the quality of the reconstruction, a total of eight sparse images were collected for each tilt angle. The downsampling amount in the sparse images ranged between 3.125% and 25%.

After collection of the whole tilt-series, block-based sparse images were reconstructed in Matlab using 250 iterations of the inpainting algorithm developed by Garcia [35,36]. Fewer iterations did not give satisfactory results whereas more iterations seemed unnecessary and cost longer computing times. Tilt-series alignment was performed using cross-correlation in Etomo [37,38] and particular care was taken to produce the best aligned tilt-series possible by playing with alignment parameters in Etomo such as filtering and trimming of the images. Reconstructions were generated by weighted-back projection (WBP) in Etomo.

2.4. Reconstruction Quality Assessment

The quality of reconstructions was initially assessed by visual inspection. However, to better characterize the reconstructions, image quality descriptors (IQD) were also measured on individual Z-slices of the 3D volumes. Entropy (H), root mean square contrast (C_{rms}) and Michelson's contrast (C_m) were computed using the following equations [39]:

$$H = -\sum_{i=1}^{n} P_i \log_2 P_i$$

H is the entropy of the element containing n pixels and P_i is the appearance probability of the pixel value i in the element.

$$C_{rms} = \sqrt{\frac{1}{MN}\sum_{i=1}^{M}\sum_{j=1}^{N}(I_{xy} - \bar{I})^2}$$

C_{rms} is the root mean square contrast, M and N are the dimensions of the data in x and y respectively, I_{xy} is the value of the pixel at position xy and $\bar{I}$ is the mean pixel value.

$$C_m = \frac{I_{max} - I_{min}}{I_{max} + I_{min}}$$

C_m represents the Michelson's contrast, I_{max} is the maximum pixel value and I_{min} is the minimum pixel value.

In ET, the Z-dimension of a reconstruction is large enough so that the entirety of the object of interest can fit in the 3D volume. The quality of the reconstructions has been estimated by computing the ratio between IQD values measured at the level of the object of interest and IQD values measured above and below where there should be no object (later on referred as the background). In an ideal missing-wedge-free and noise-free 3D reconstruction, Z-slices located above and below the object of interest should have no contrast. In a 3D reconstruction based on experimental images that contain noise, the same Z-slices suffer from missing-wedge and noise reconstruction artifacts.

2.5. Data Presentation

Images presented in this work have been generated using ImageJ (v1.51j8) [40]. Artworks have been designed in Blender. The computation of IQD and corresponding figures were performed in Matlab. IQD ratios data were fitted to the following model function using non-linear least-squares regression:

$$f(x) = A - 1/(B + Cx)$$

3. Results

3.1. Practical Implementation in 2D

3.1.1. Collection of Sparse STEM Images

Collection of non-overlapping pixels blocks randomly distributed over the area of interest (ROI) was performed using an in-house developed Digital Micrograph (DM) script. The Digiscan II software of DM has built-in functions that allow the collection of pixel blocks which dimensions can be 16 × 4 pixels. This collection scheme was used on a carbon crossed line grating grid to collect 12.5% of the pixels (Figure 1A). The white background represents uncollected pixels. Inpainting was then computed to recover missing pixels (Figure 1B). The pattern of the crossed line grating replica becomes apparent after the inpainting reconstruction and resembles the ground-truth image of the same ROI (Figure 1C).

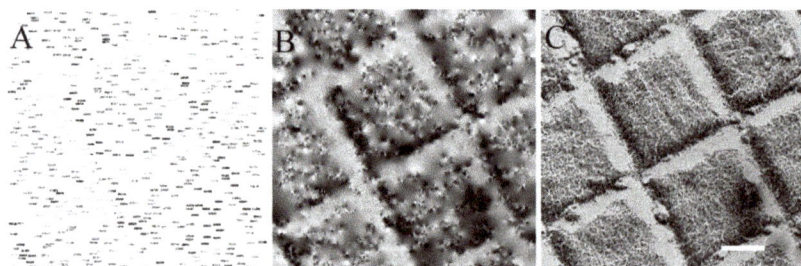

Figure 1. Collection of a sparse scanning transmission electron microscopy (STEM) image, inpainting reconstruction and comparison with ground truth. (**A**) Sparse STEM image downsampled to 12.5% using 512 blocks of 16 × 4 pixels. The white background corresponds to uncollected pixels. (**B**) Sparse image A after inpainting reconstruction. (**C**) Ground truth fully-collected STEM image of the very same ROI on the grid. Scale bar is 200 nm. The three subfigures have the same scale.

3.1.2. Accuracy of the Electron Beam Positioning during Sparse Imaging

In the work of Anderson et al., a similar strategy was attempted to perform non-linear scanning using the scanning coils of a scanning electron microscope and the authors pointed out the weak accuracy of the beam positioning in such non-trivial beam motions [27]. During the first tests, there was no apparent issue with the accuracy of the beam positioning on the JEOL 2200FS since the method reconstituted well the crossed line grating pattern (Figure 1). To further evaluate the accuracy of the beam positioning, the methodology was tested at much higher resolution with a second electron microscope, a CS-corrected JEOL 2100F. High-resolution images of the crystal structure of an iron oxide (FeO) nanoparticle [32] were obtained with that microscope (Figure 2). Overlapping blocks were collected so that positioning inaccuracies would be evidenced by the disruption of the crystalline lattice. Overlapping blocks are easily recognized and it is possible to focus on a region where the crystalline structure is visible over several overlapping blocks (black box in Figure 2A). Four thick grey lines have been drawn along two different axes of the crystal lattice passing through several blocks (Figure 2B). The atom columns are well aligned with the thick grey lines and no disruption of the crystalline lattice is observed. To further characterize the beam positioning, the distance between atom columns has been systematically measured and do not show visible discrepancy between collected blocks (Figure S1). The stability of the electron beam has been tested and verified with two different electron microscopes at different ranges of magnifications, validating the strategy to use Digiscan II and DM to drive the electron beam using the scanning coils.

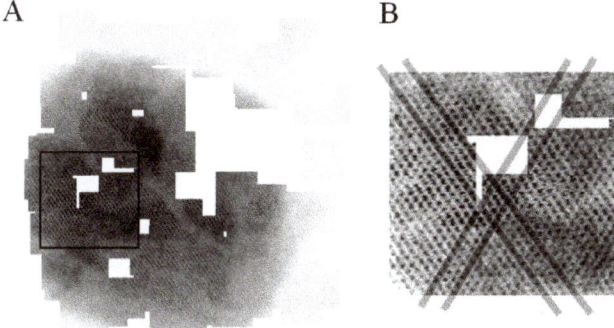

Figure 2. Block-based sparse imaging of an FeO nanoparticle at very high magnification. (**A**) Overlapping blocks (64 × 64 pixels) partially covering the FeO nanoparticle, white areas correspond to uncollected zones. (**B**) Zoom-in on a particular location of the nanoparticle (corresponding to the black box in (**A**)) where the crystalline pattern is visible on several overlapping blocks. Thick grey lines drawn along the crystalline lattice highlight the alignment of the atom columns over several blocks.

3.2. Practical Implementation in 3D

The block-based sparse imaging and the reconstruction of the missing information using inpainting performed well on the crossed line grating replica. This sample has the advantage of being electron-resistant and possesses interesting features such as the Au/Pd grains and spherical latex beads, it is a sample of choice to develop the method for tomographic collection of sparse STEM images. The method relies on the acquisition of a tilt-series composed of sparse STEM images that are subsequently reconstructed using inpainting, then aligned by cross-correlation and eventually 3D reconstructed using WBP. The sparse 3D data are compared to a ground-truth reference made of fully-collected images.

3.2.1. Collection of Sparse STEM Tilt-Series

In a tomography workflow, there are two image processing steps that need to be performed accurately during the data collection: (i) the image registration to track the sample and (ii) the measurement of the focus value. Regarding the image registration step, the shifts between two consecutive tilts have to be computed to correct the sample drift. Regarding the focusing step, in STEM, it is usually performed by finding the most contrasted image among several images collected in a range of focus values. Sparse images might not have sufficient information in common to accurately perform image registration or contrast measurement. To ease the two image processing steps as mentioned above, they are performed on fully-collected images instead of sparse ones. Since the method aims to reduce the electron dose in the ROI, a second region next to the ROI and aligned with the tilt axis is used to acquire the fully-collected images to perform both focusing and tracking tasks (Figure 3). These areas are later on referred as focusing and tracking areas. The whole tomography workflow has been scripted and developed in DM. The script also codes for a graphical user interface to input the data collection parameters (Figure S2). The main steps of the script are the following:

- **Step 1—Focusing**: determination of the 0 nm focus value by successive acquisitions of fully-collected images on the focusing area in a focus range.
- **Step 2—Tracking**: acquisition of a fully-collected image on the tracking area and determination of the drift that occurred compared to the prior tilt angle by cross-correlation.
- **Step 3—Sparse data collection**: acquisition of the block-based sparse image on the region of interest.
- **Step 4—End of current tilt**: rotation to the next tilt angle and back to step 1.

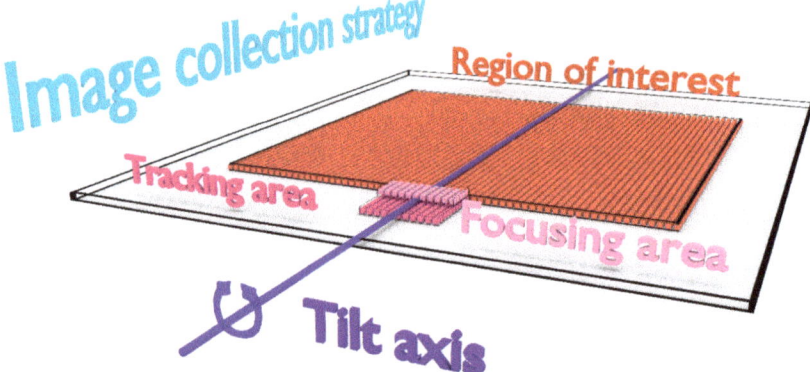

Figure 3. Strategy developed to collect sparse STEM images in a tomography workflow. The region of interest (in red) and the tracking and focusing areas (in dark and light pink respectively) are aligned with the tilt axis (in dark blue).

3.2.2. Validation of the Sparse STEM Tomography Workflow

For validation, the 3D reconstruction computed on sparse images has been compared with a reference volume. To this purpose, the sparse tomography workflow (6.25% downsampling) has been compared to a reference collection scheme (i.e., made of fully-collected images). Sparse images and fully-collected ones were consecutively collected at the same tilt-angle on the same ROI of a crossed line grating grid so that both images represent the object in the same orientation. After collection of the sparse data, missing pixels were first reconstructed using inpainting. Secondly, both tilt-series (i.e., sparse and reference ones) were aligned using the cross-correlation alignment parameters computed on the inpainted sparse images so that the quality of the alignment performed on inpainted data can be estimated. Finally, both sparse and reference 3D reconstructions were computed using WPB in Etomo (Figure 4). Z-slices 50 of the reconstructions (Figure 4, left column) do not contain any sample, however contrasted structures appear (Figure 4, white asterisks). These structures correspond to reconstruction artifacts indicating that the alignment could be improved. It can be noted that reconstruction artifacts in the sparse reconstruction are more visible than the ones in the reference reconstruction, most probably since sparse images originally contain fewer collected pixels (6.25% downsampling). At the level of the sample (Z-slices 150) the reference reconstruction is more contrasted and more detailed than the sparse one (Figure 4, center column). Individual Au/Pd deposits can be discriminated in the reference reconstruction whereas only global shapes of deposit clusters are visible in the sparse reconstruction (Figure 4, zoom-in). At the level of the latex spheres (Z-slices 250), similar reconstruction artifacts as the ones observed on Z-slices 50 are present (Figure 4, right column). Latex spheres (Figure 4, white arrows) are easily recognized in the sparse reconstruction and their roundness is conserved, though less defined than the ones in the reference reconstruction.

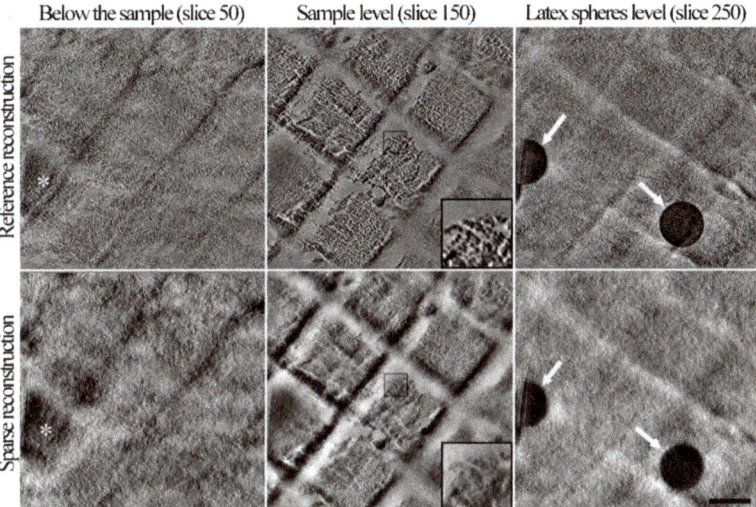

Figure 4. Comparison of the reference and sparse reconstructions. The reference reconstruction (**upper row**) has been computed using fully-collected images whereas the sparse reconstruction (**lower row**) has been computed using sparse images containing 6.25% of the pixels. For each reconstructed volume, three Z-slices were extracted: (i) below the sample (left column, Z-slices 50), (ii) at the level of the Au/Pd coating (center column, Z-slices 150) and (iii) above the sample at the level of the latex spheres (right column, Z-slices 250). The center area of Z-slices 150 (black boxes) has been zoomed-in to better visualize the details on the Au/Pd deposits. Reconstructions artifacts and latex spheres are indicated using white asterisks and white arrows, respectively. The scale bar is 250 nm. The six subfigures have the same scale.

The 3D reconstruction computed from sparse images contains enough details to describe the overall structure of the carbon replica grid, the clusters of Au/Pd grains and latex spheres are visible. However, at such downsampling, the sparse reconstruction is not as detailed as the reference one, most probably because the very low amount of collected pixels cannot be compensated by the inpainting treatment. The presence of reconstruction artifacts showed that the cross-correlation alignment on inpainted sparse images could be improved. Potentially, better alignment could be computed if images of higher quality were used (i.e., if more pixels were collected). The carbon replica grid has been used as a test sample because it has several advantages. First, it is a thin sample that theoretically necessitates fewer projections to be accurately reconstructed in 3D compared to a thick sample. Secondly, its simple composition (low atomic variety) is theoretically better described using heavily downsampled sparse images compared to complex samples such as biological ones. The considerable downsampling (6.25%)

used on the test sample authorizes some room to adapt the method to other types of sample if higher amounts of pixels need to be collected. These aspects are discussed in the following part that focuses on the application of the method on a 500 nm-thick section of resin-embedded *T. brucei* sample.

3.3. Application on a Biological Sample

The 500 nm-thick section of resin-embedded *T. brucei* represents the typical kind of sample that is studied in ET by life scientists. Eight tilt-series constituted of sparse images which downsampling ranged between 3.125% and 25% were collected on the 500 nm-thick resin section. Several downsampling values were tested to verify how low sparse imaging can be diminished while maintaining sufficient structural details. After data collection, tilt-series were inpainted, aligned using cross-correlation and reconstructed using WPB, giving rise to eight different volumes containing the ROI of the *T. brucei* section. The quality of the reconstructions has been assessed both visually (Figure 5) and using IQD (Figure 6). Displayed images represent the central slices of the reconstructions (Figure 5). The center of the ROI contains the flagellar pocket of a flagellum (Figure 5, Mt) and the nucleus of the cell is present in the top right corner (Figure 5, N). Visually, the amount of structural details in the reconstructions does not seem linear with the amount of collected pixels. At 3.125% downsampling, the reconstruction suffers important blurring and cellular structures are hardly recognizable. If the amount of collected pixels is doubled (6.25%) or tripled (9.375%), the increase in details is important and structures in the cell cytoplasm start to arise from the blurry background (Figure 5, Cyt). At 12.5% downsampling the structural information is more detailed, the membranes (Figure 5, Mb) are more continuous and the microtubule doublets (Figure 5, Mt) of the flagellum are better defined. Around the nucleus (Figure 5, N), the two membranes of the nuclear envelope (Figure 5, Ne) can be resolved. By eye, differences between reconstructions ranging from 15.625% to 25% downsampling are thin despite the important increase of collected pixels (e.g., there are 62.5% more pixels collected in the 25% downsampling compared to the 15.625% one). Plot profiles passing through known structures of the flagellar pocket of *T. brucei* were computed to better display the variation of pixel intensity in the various reconstructions (Figure S3).

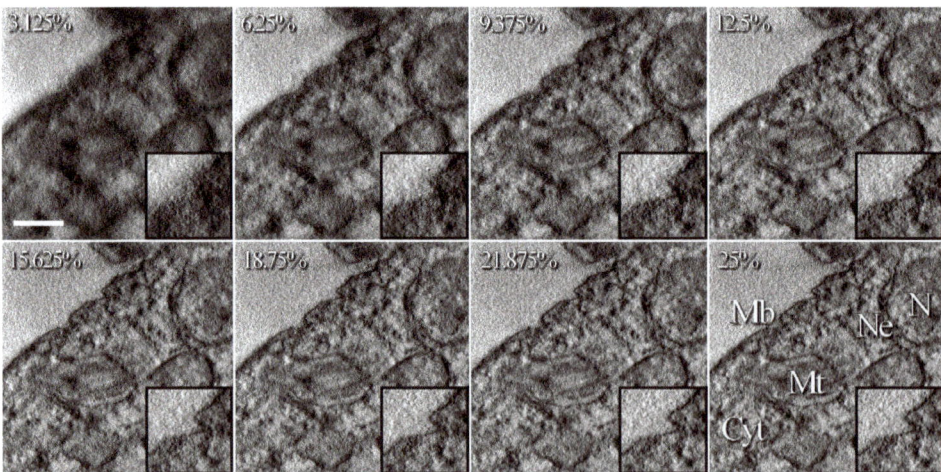

Figure 5. Comparison of 3D volumes reconstructed from sparse images collected at different downsampling values. The central Z-slice of each reconstruction is displayed. The value in the top left corner of each image corresponds to the downsampling value. Inserts in the bottom right corner are zoom-ins of the cell membrane. Several cellular structures are pointed out: the cytoplasm (Cyt), microtubules of the flagellum (Mt), the cell membrane (Mb), the nucleus (N) and the nuclear envelope (Ne). The scale bar is 400 nm. The eight subfigures have the same scale.

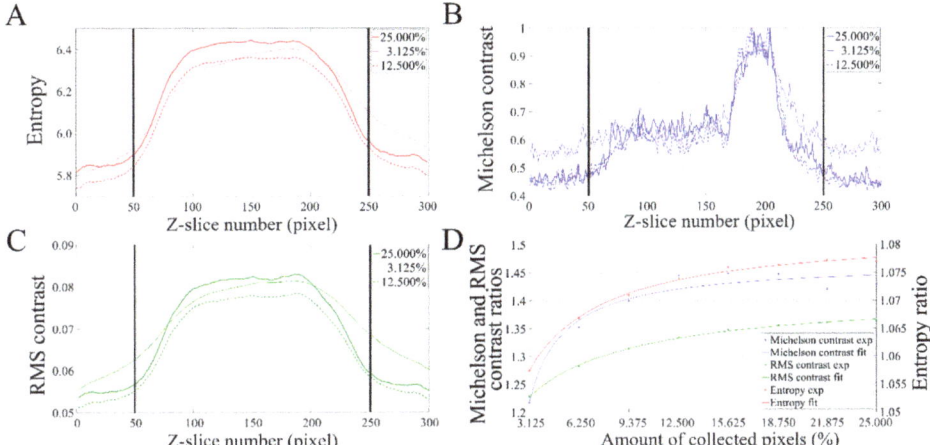

Figure 6. Image quality descriptors (IQD) values of sparse reconstructions depending on the amount of downsampling. (**A**–**C**) Plots of the H, C_m and C_{rms} descriptors computed at each Z slice of three reconstructions (3.125%, 12.5% and 25%), respectively. Thick vertical lines mark the location of the resin section in the reconstructed volumes (between Z-slices 50 and 250). (**D**) IQD ratios (experimental values and fitted curves) plotted by increasing order of downsampling.

To further characterize the reconstructions and to better describe thin differences that were not discernible by visual inspection of the volumes, IQD values were computed on individual Z-slices of all reconstructions. For the sake of clarity, because of the high number of reconstructions, only IQD values of some reconstructions (i.e., 3.125%, 12.5% and 25%) are presented (Figure 6A–C). Then, ratios between the average of IQD values of Z-slices passing through the resin section (i.e., between Z-slices 50 and 250) and the average of IQD values of Z-slices not passing through the resin section (i.e., the background, between Z-slices 1 and 50 and between Z-slices 250 and 300) have been computed and are used to evaluate the quality of the reconstructions (see Section 2.4). Ratios are presented and have been plotted by increasing order of downsampling value (Figure 6D).

For all reconstructions, IQD values at the level of the sample (i.e., between Z-slices 75 and 225) are greater than that of Z-slices where there is no sample (Figure 6A–C). Note that the high C_m values around Z-slices 200 correspond to the presence of heavily contrasted structures in the reconstructions (Figure 6B). C_m values being very sensitive to variations of pixels intensities because of the way it is computed, as displayed by its saw teeth shape. At 3.125% downsampling, when crossing from the resin section to the background (and vice versa), the slope of the IQD plots is not steep indicating that at very low amounts of collected pixels, the contrast of the object is poor (Figure 6A–C, continuous thin plots). At 12.5% downsampling, IQD plots have steeper slopes, allowing a clear discrimination between the object and the background (Figure 6A–C, dashed thick plots). At 25% downsampling, IQD values have similar behavior as that of 12.5% downsampling (Figure 6A–C, continuous thick plots). The ratio of IQD values has been computed to take into account the background noise so that reconstructions can be compared (Figure 6D). Ratios of IQD values greatly increase between 3.125% and 12.5% downsampling. At 3.125% downsampling, C_m, C_{rms} and H ratios are 1.2182, 1.2289 and 1.0575, respectively. At 12.5% downsampling, C_m, C_{rms} and H ratios are 1.4455, 1.3347 and 1.0740 respectively. At 25% downsampling, C_m, C_{rms} and H ratios reach 1.4296, 1.3650 and 1.0769 respectively. As can be seen with the fitted curves, above 12.5% downsampling, IQD ratios do not increase much and they start reaching a plateau around 25% downsampling. Based on the curve fitting, if all pixels were collected, ratios obtained for C_m, C_{rms} and H would be 1.50, 1.39 and 1.08 respectively. A table summarizing values of IQD ratios is available in Supplementary Information (Table S1). The three downsampling ratios presented in Figure 6A–C were chosen since they are good descriptors of the evolution of IQD values depending on the downsampling. The combination of (i) the three IQD measurements, (ii) the steepness of IQD plots at transitions between resin section and background and (iii) IQD ratios computed between the resin section and the background allows the characterization of thin differences that exist between the various reconstructions. The numerical IQD measurements (Figure 6) are in agreement with the visual inspection of the reconstructions (Figure 5).

4. Discussion

The aim of this work was to develop a tomography workflow to collect sparse images, only relying on the basic equipment of STEM electron microscopes to drive the electron beam (i.e., the scanning coils).

In the literature, scanning coils have previously been used to drive the electron beam in non-linear motions at sub-microsecond speeds. Anderson et al. used a dedicated system to send commands, receive detector signals and calibrate the mismatch between the desired beam position and the actual one since beam dynamics are important when high speed is achieved and when non-trivial motions are performed [27]. In the present study, the first step was to characterize how well the scanning coils of a JEOL 2200FS performed when they are commanded by Digiscan II. The first tests performed at relatively low magnification on the crossed line grating grid confirmed the correct positioning of the beam (Section 3.1.1). To push further the characterization, additional tests were performed on a CS-corrected JEOL 2100F at much higher magnification (Section 3.1.2). A good accuracy of the beam positioning was obtained and it confirmed that the strategy could be used on at least two different models of electron microscopes of the manufacturer JEOL. Somewhat, this high accuracy does not agree with what has previously been mentioned about beam dynamics [27]. It is possible that the scanning speed employed in the present study (between 1 µs and 8 µs per pixel) is slow enough not to introduce enough beam dynamics to produce a visible effect, even at high magnification. Furthermore, DM sets the block scan limit to 16 × 4 pixels and does not give the possibility to scan single pixels. This limit could exist to prevent beam dynamics from occurring. Such information could not be confirmed neither by people from Gatan USA nor Gatan France. Moreover, SEM studies are made at relatively low magnifications compared to TEM ones, hence large surfaces have to be scanned in SEM and the positioning of the beam might not be accurate when it is subjected to important deflections.

The collection time of a STEM image is defined by the number of collected pixels per line multiplied by the dwell time (time spent per pixel) and eventually by the number of lines in the image. When the beam reaches the end of a line, it is repositioned at the beginning of the next line (this step is called fly-back). Usually, the fly back time (about 150 to 250 µs) is relatively small compared to the time spent to collect the pixels of a single line (several thousands of µs). However, when scanning 16 × 4 pixel blocks using the scanning coils, a line is made of 16 pixels only and the fly-back command is called every 16 pixels. Eventually, the fly-back command is called so often that it contributes to a substantial part of the whole collection time. Hopefully, pixel blocks do not cover the whole ROI so the collection time stays reasonable. In practice, the whole acquisition of a tilt-series constituted of sparse images with 10% of the pixels, collected between −60° and +60° using 2° increments take about 220 min to complete. In the setups of Béché et al. and Vanrompay et al., a fast electromagnetic shutter was installed next to the condenser of the electron microscope so that they were able to blank the electron beam using custom sequences while the scanning coils performed standard linear acquisition [26,29]. In the setup of Anderson et al., the beam was scanned at very fast speeds (about 400 ns/pixel) using a custom scanning system [27]. Interestingly, these setups allowed the collection of sparse images at least as fast as standard linearly acquired images. Instead of collecting blocks, the acquisition of lines would be interesting to speed up the collection of sparse images using the scanning coils of the electron microscope as it is performed in Li et al. [28]. However, since lines cannot fill a ROI as efficiently as small blocks, with the exception maybe of Lissajous scans, similar downsampling values might not be reached.

Regarding the processing of the sparse images, several inpainting algorithms were tried and the one developed by Garcia [35,36] gave the best results in terms of visual quality. The inpainted sparse images do not have sufficient details to use fiducial-based [38] or landmark-based [41] alignment methods; hence, image registration was then performed by cross-correlation means. Reconstruction artifacts most probably originating from a perfectible alignment were observed on the crossed lined grating sample (Section 3.2.2) but not on the *T. brucei* sample (Section 3.3). Images of the crossed line grating sample had few latex spheres that had a strong contrast compared to the carbon where Au/Pd grains were deposited at the surface. Cross-correlation might fail in producing a good alignment on such images. On the contrary, images of the *T. brucei* sample were collected on a ROI that contained several cellular elements with strong contrast. The results show that the quality of the alignment is good enough to describe cellular structures in the reconstructed volume of a 500 nm-thick resin section. Testing other reconstruction algorithms such as iterative reconstruction methods or compressive-sensing approaches to verify which algorithm reconstructs the best such sparsely collected data is out of the scope of this work. Each reconstruction solution would need to be adapted to this specific kind of data.

If higher resolutions are required, cross-correlation solutions might not be sufficient to align the images with enough accuracy and other or new alignment methods should be employed, the main difficulty lying in the fact that the images of the tilt-series share very little amount of common information. One solution could be to generate an initial cross-correlation-based 3D reconstruction which alignment could be improved by iterative reconstruction algorithms that refine the alignment during the projection/back-projection comparison step, as described in previous studies [42,43]. Moreover, it seems necessary to design an algorithm that can discriminate between collected pixels (which intensity values have been measured experimentally) from inpainted ones (which intensity values have been estimated computationally). Experimental intensity values should have more weight in the reconstruction compared to estimated ones. Combining more robust image reconstruction algorithms such as wavelet- or shearlet-based inpainting and more accurate tilt-series alignment methods should help improving the resolution at levels comparable to resolutions achieved in classical ET.

Using the proposed approach; sparse STEM images containing only 15% of the pixels were collected on a 500 nm-thick resin section of *T. brucei*. The total electron dose received by the sample after the tilt-series collection was about 10 e$^-$/Å^2. The resulting 3D reconstruction contained enough structural details to recognize typical components present in the cytoplasm of eukaryotic cell. The method could be used both in material and life sciences to diminish the electron dose. In material sciences, similar strategies could be applied in 2D or in 3D to study in situ fragile samples such as nanowires [44] or Li-Ion batteries [45]. In life sciences, it could be used to study beam-sensitive cryo-samples in ET or in correlative experiments where the sample is exposed to different kinds of radiations. Correlative approaches are new powerful investigation methods and acquisition protocols limiting radiation damages could contribute to the development of correlative approaches yet to be proposed.

Supplementary Materials: The following are available online at http://www.mdpi.com/1996-1944/12/14/2281/s1, Figure S1: Intra-block and inter-block distances between atom columns. (A) Image showing the location of blocks that have been collected on the FeO nanoparticle. Each block has a unique grey value to better discriminate them. The darker the block, the earlier (i.e., in time) it has been collected, hence the whiter the block, the later it has been collected. (B) Diagram presenting the locations where the distance between atom columns have been computed. Four thick grey lines highlight the precise locations. Distances have been measured on two different axes of the crystalline pattern not to miss a drift that would occur in one direction only, for more accurate error determination. (C) Close-up view showing the various superimposed blocks. It is possible to determine if two consecutive atom columns belong to the same collected block or if they belong to two different blocks. (D) Plot showing the measured distance between consecutive atom columns. Atom columns belonging to the same block (intra-block, blue and red points) and atom columns belonging to two different blocks (inter-block, blue and red circles) have been plotted separately. Blue and red plots represent values measured on axis1 and axis2 respectively. Mean distance values for axis1 (intra-block), axis1 (inter-block), axis2 (intra-block) and axis2 (inter-block) are 5.47 (std: 0.96), 6.22 (std: 1.09), 6.24 (std: 1.223), 5.64 (std: 0.86) pixels respectively. Intra-block values show variations originating most probably from noise and crystalline defects in the nanoparticle. Inter-block measurements have similar values, demonstrating that the beam positioning does not introduce disruption of the crystalline lattice, proving that the method is accurate enough to perform sparse imaging using pixel blocks. Figure S2: Graphical interface of the Digital Micrograph script to perform sparse STEM tomography. Standard parameters such as the microscope magnification, the dwell time, the angular range and the tilt-step are set up in the interface. Parameters specific to the sparse imaging are present: scan (block) size (e.g., 16 × 4 px), the reconstructed image size (e.g., 2000 px) and the number of collected scans (blocks) (e.g., 512). The focus value parameter serves as an initial value around which the script will search for the 0 nm defocus value. Figure S3: Intensity profiles of the same ROI in 3D volumes reconstructed from variously downsampled sparse STEM images. The profiles have taken through the flagellum of *T. brucei* at the position indicated by the pointed yellow line on the tomographic slice. The height of the intensity profiles does not inform about the intensity values, they have been placed on top of each other for the sake of clarity. The intensity profiles cross several structures of the cell, the cytoplasm (Cyt), the flagellar pocket membrane (FPM), the flagellar pocket space (FPS), the flagellar membrane (FM), the transition zone space (TZS), the microtubule doublets of the axoneme (Mt) and the axonemal space (AS). If downsampling is lower than about 10%, intensity variation from one structure to the other is hardly recognizable. When greater amounts of pixels are collected, strong intensity variations allow to discriminate the structures. Above 15% downsampling the intensity profiles are very similar one another. Table S1: Table summarizing values of IQD ratios for three downsampling (3.125%, 12.5% and 25%) and for the estimated 100% corresponding to no downsampling. The first row contains the downsampling values. The last column represents the values extrapolated from the fitted curves. Values in between brackets represent the ratio with respect to the extrapolated values.

Funding: This research was funded by two ANR grants (ANR-11-BSV8-016 and ANR-15-CE11-0002).

Acknowledgments: The author is greatly indebted to G. Melinte and O. Ersen for providing access to the CS-corrected JEOL 2100F at IPCMS (Strasbourg, France) to test the block-based sparse imaging at high-resolution on FeO nanoparticles. The author acknowledges the laboratory of P. Bastin at Institut Pasteur (Paris, France) for providing the *T. brucei* sample. The author acknowledges M. Ribardière and J.-P. Dérouet from JEOL France SAS (Croissy-sur-Seine, France) for the multiple constructive discussions about scanning process in JEOL electron microscopes. J.-P. Michel at Institut Galien (Châtenay-Malabry, France) is acknowledged for his critical reading of the manuscript. The author acknowledges the PICT-IBiSA for providing access to the cryo-electron microscopy facility.

Conflicts of Interest: The author declares no conflict of interest.

References

1. Lacomble, S.; Vaughan, S.; Gadelha, C.; Morphew, M.K.; Shaw, M.K.; McIntosh, J.R.; Gull, K. Basal body movements orchestrate membrane organelle division and cell morphogenesis in Trypanosoma brucei. *J. Cell Sci.* **2010**, *123*, 2884–2891. [CrossRef] [PubMed]
2. Lucic, V.; Rigort, A.; Baumeister, W. Cryo-electron tomography: The challenge of doing structural biology in situ. *J. Cell Biol.* **2013**, *202*, 407–419. [CrossRef] [PubMed]
3. Gan, L.; Jensen, G.J. Electron tomography of cells. *Q. Rev. Biophys.* **2012**, *45*, 27–56. [CrossRef] [PubMed]
4. Oikonomou, C.M.; Jensen, G.J. A new view into prokaryotic cell biology from electron cryotomography. *Nat. Rev. Microbiol.* **2017**, *15*, 128. [CrossRef] [PubMed]
5. Schaffer, M.; Engel, B.D.; Laugks, T.; Mahamid, J.; Plitzko, J.M.; Baumeister, W. Cryo-focused Ion Beam Sample Preparation for Imaging Vitreous Cells by Cryo-electron Tomography. *Bio-Protocol* **2015**, *5*, 1575. [CrossRef]
6. Mahamid, J.; Pfeffer, S.; Schaffer, M.; Villa, E.; Danev, R.; Cuellar, L.K.; Forster, F.; Hyman, A.A.; Plitzko, J.M.; Baumeister, W. Visualizing the molecular sociology at the HeLa cell nuclear periphery. *Science* **2016**, *351*, 969–972. [CrossRef]
7. Bertiaux, E.; Mallet, A.; Fort, C.; Blisnick, T.; Bonnefoy, S.; Jung, J.; Lemos, M.; Marco, S.; Vaughan, S.; Trepout, S.; et al. Bidirectional intraflagellar transport is restricted to two sets of microtubule doublets in the trypanosome flagellum. *J. Cell Biol.* **2018**, *217*, 4284–4297. [CrossRef]
8. Soto, G.E.; Young, S.J.; Martone, M.E.; Deerinck, T.J.; Lamont, S.; Carragher, B.O.; Hama, K.; Ellisman, M.H. Serial section electron tomography: A method for three-dimensional reconstruction of large structures. *NeuroImage* **1994**, *1*, 230–243. [CrossRef]
9. Midgley, P.A.; Weyland, M. 3D electron microscopy in the physical sciences: The development of Z-contrast and EFTEM tomography. *Ultramicroscopy* **2003**, *96*, 413–431. [CrossRef]
10. Pennycook, S.J.; Nellist, P.D. *Scanning Transmission Electron Microscopy*; Springer: New York, NY, USA, 2011.
11. Hohmann-Marriott, M.F.; Sousa, A.A.; Azari, A.A.; Glushakova, S.; Zhang, G.; Zimmerberg, J.; Leapman, R.D. Nanoscale 3D cellular imaging by axial scanning transmission electron tomography. *Nat. Methods* **2009**, *6*, 729–731. [CrossRef]
12. Aoyama, K.; Takagi, T.; Hirase, A.; Miyazawa, A. STEM tomography for thick biological specimens. *Ultramicroscopy* **2008**, *109*, 70–80. [CrossRef] [PubMed]

13. Sousa, A.A.; Hohmann-Marriott, M.F.; Zhang, G.; Leapman, R.D. Monte Carlo electron-trajectory simulations in bright-field and dark-field STEM: Implications for tomography of thick biological sections. *Ultramicroscopy* **2009**, *109*, 213–221. [CrossRef] [PubMed]
14. Sousa, A.A.; Azari, A.A.; Zhang, G.; Leapman, R.D. Dual-axis electron tomography of biological specimens: Extending the limits of specimen thickness with bright-field STEM imaging. *J. Struct. Biol.* **2011**, *174*, 107–114. [CrossRef] [PubMed]
15. Wolf, S.G.; Houben, L.; Elbaum, M. Cryo-scanning transmission electron tomography of vitrified cells. *Nat. Methods* **2014**, *11*, 423–428. [CrossRef] [PubMed]
16. Rez, P.; Larsen, T.; Elbaum, M. Exploring the theoretical basis and limitations of cryo-STEM tomography for thick biological specimens. *J. Struct. Biol.* **2016**, *196*, 466–478. [CrossRef] [PubMed]
17. Batenburg, K.J.; Bals, S.; Sijbers, J.; Kubel, C.; Midgley, P.A.; Hernandez, J.C.; Kaiser, U.; Encina, E.R.; Coronado, E.A.; Van Tendeloo, G. 3D imaging of nanomaterials by discrete tomography. *Ultramicroscopy* **2009**, *109*, 730–740. [CrossRef] [PubMed]
18. Goris, B.; Van den Broek, W.; Batenburg, K.J.; Heidari Mezerji, H.; Bals, S. Electron tomography based on a total variation minimization reconstruction technique. *Ultramicroscopy* **2012**, *113*, 120–130. [CrossRef]
19. Saghi, Z.; Holland, D.J.; Leary, R.; Falqui, A.; Bertoni, G.; Sederman, A.J.; Gladden, L.F.; Midgley, P.A. Three-dimensional morphology of iron oxide nanoparticles with reactive concave surfaces. A compressed sensing-electron tomography (CS-ET) approach. *Nano Lett.* **2011**, *11*, 4666–4673. [CrossRef] [PubMed]
20. Saghi, Z.; Divitini, G.; Winter, B.; Leary, R.; Spiecker, E.; Ducati, C.; Midgley, P.A. Compressed sensing electron tomography of needle-shaped biological specimens–Potential for improved reconstruction fidelity with reduced dose. *Ultramicroscopy* **2016**, *160*, 230–238. [CrossRef] [PubMed]
21. Donati, L.; Nilchian, M.; Trepout, S.; Messaoudi, C.; Marco, S.; Unser, M. Compressed sensing for STEM tomography. *Ultramicroscopy* **2017**, *179*, 47–56. [CrossRef] [PubMed]
22. Jiang, Y.; Padgett, E.; Hovden, R.; Muller, D.A. Sampling limits for electron tomography with sparsity-exploiting reconstructions. *Ultramicroscopy* **2018**, *186*, 94–103. [CrossRef] [PubMed]
23. Monier, E.; Oberlin, T.; Brun, N.; Tencé, M.; de Frutos, M.; Dobigeon, N. Reconstruction of partially sampled multi-band images—Application to STEM-EELS imaging. *IEEE Trans. Comput. Imaging* **2018**, *4*, 585–598. [CrossRef]
24. Stevens, A.; Yang, H.; Carin, L.; Arslan, I.; Browning, N.D. The potential for Bayesian compressive sensing to significantly reduce electron dose in high-resolution STEM images. *Microscopy* **2014**, *63*, 41–51. [CrossRef] [PubMed]
25. Saghi, Z.; Benning, M.; Leary, R.; Macias-Montero, M.; Borras, A.; Midgley, P.A. Reduced-dose and high-speed acquisition strategies for multi-dimensional electron microscopy. *Adv. Struct. Chem. Imaging* **2015**, *1*, 7. [CrossRef]
26. Béché, A.; Goris, B.; Freitag, B.; Verbeeck, J. Development of a fast electromagnetic shutter for compressive sensing imaging in scanning transmission electron microscopy. *Appl. Phys. Lett.* **2016**, *108*, 093103. [CrossRef]
27. Anderson, H.S.; Ilic-Helms, J.; Rohrer, B.; Wheeler, J.; Larson, K. Sparse Imaging for Fast Electron Microscopy. In Proceedings of the Computational Imaging XI. International Society for Optics and Photonics, Burlingame, CA, USA, 14 February 2013; Volume 8657, p. 86570C. [CrossRef]
28. Li, X.; Dyck, O.; Kalinin, S.V.; Jesse, S. Compressed Sensing of Scanning Transmission Electron Microscopy (STEM) With Nonrectangular Scans. *Microsc. Microanal.* **2018**, *24*, 623–633. [CrossRef] [PubMed]
29. Vanrompay, H.; Béché, A.; Verbeeck, J.; Bals, S. Experimental Evaluation of Undersampling Schemes for Electron Tomography of Nanoparticles. *Part. Part. Syst. Charact.* **2019**, 1900096. [CrossRef]
30. Ferroni, M.; Signoroni, A.; Sanzogni, A.; Masini, L.; Migliori, A.; Ortolani, L.; Pezza, A.; Morandi, V. Biological application of Compressed Sensing Tomography in the Scanning Electron Microscope. *Sci. Rep.* **2016**, *6*, 33354. [CrossRef] [PubMed]
31. Guay, M.D.; Czaja, W.; Aronova, M.A.; Leapman, R.D. Compressed Sensing Electron Tomography for Determining Biological Structure. *Sci. Rep.* **2016**, *6*, 27614. [CrossRef]
32. Pichon, B.P.; Gerber, O.; Lefevre, C.; Florea, I.; Fleutot, I.; Baaziz, W.; Pauly, M.; Ohlmann, M.; Ulhaq, C.; Ersen, O.; et al. Microstructural and Magnetic Investigations of Wüstite-Spinel Core-Shell Cubic-Shaped Nanoparticles. *Chem. Mater.* **2011**, *23*, 2886–2900. [CrossRef]
33. Brun, R.; Schonenberger. Cultivation and in vitro cloning or procyclic culture forms of Trypanosoma brucei in a semi-defined medium. Short communication. *Acta Trop.* **1979**, *36*, 289–292. [PubMed]

34. Trépout, S.; Tassin, A.M.; Marco, S.; Bastin, P. STEM tomography analysis of the trypanosome transition zone. *J. Struct. Biol.* **2018**, *202*, 51–60. [CrossRef] [PubMed]
35. Garcia, D. Robust smoothing of gridded data in one and higher dimensions with missing values. *Comput. Stat. Data Anal.* **2010**, *54*, 1167–1178. [CrossRef] [PubMed]
36. Wang, G.; Garcia, D.; Liu, Y.; de Jeu, R.; Dolman, A.J. A three-dimensional gap filling method for large geophysical datasets: Application to global satellite soil moisture observations. *Environ. Model. Softw.* **2012**, *30*, 139–142. [CrossRef]
37. Kremer, J.R.; Mastronarde, D.N.; McIntosh, J.R. Computer visualization of three-dimensional image data using IMOD. *J. Struct. Biol.* **1996**, *116*, 71–76. [CrossRef]
38. Mastronarde, D.N.; Held, S.R. Automated tilt series alignment and tomographic reconstruction in IMOD. *J. Struct. Biol.* **2017**, *197*, 102–113. [CrossRef] [PubMed]
39. Peli, E. Contrast in complex images. *J. Opt. Soc. Am. A Opt. Image Sci.* **1990**, *7*, 2032–2040. [CrossRef]
40. Schneider, C.A.; Rasband, W.S.; Eliceiri, K.W. NIH Image to ImageJ: 25 years of image analysis. *Nat. Methods* **2012**, *9*, 671–675. [CrossRef] [PubMed]
41. Sorzano, C.O.; Messaoudi, C.; Eibauer, M.; Bilbao-Castro, J.R.; Hegerl, R.; Nickell, S.; Marco, S.; Carazo, J.M. Marker-free image registration of electron tomography tilt-series. *BMC Bioinform.* **2009**, *10*, 124. [CrossRef] [PubMed]
42. Yu, L.; Snapp, R.R.; Ruiz, T.; Radermacher, M. Projection-based volume alignment. *J. Struct. Biol.* **2013**, *182*, 93–105. [CrossRef]
43. Tomonaga, S.; Baba, M.; Baba, N. Alternative automatic alignment method for specimen tilt-series images based on back-projected volume data cross-correlations. *Microscopy* **2014**, *63*, 279–294. [CrossRef] [PubMed]
44. Zhang, C.; Kvashnin, D.G.; Bourgeois, L.; Fernando, J.F.S.; Firestein, K.; Sorokin, P.B.; Fukata, N.; Golberg, D. Mechanical, Electrical, and Crystallographic Property Dynamics of Bent and Strained Ge/Si Core–Shell Nanowires As Revealed by in situ Transmission Electron Microscopy. *Nano Lett.* **2018**, *18*, 7238–7246. [CrossRef] [PubMed]
45. Karakulina, O.M.; Demortière, A.; Dachraoui, W.; Abakumov, A.M.; Hadermann, J. In Situ Electron Diffraction Tomography Using a Liquid-Electrochemical Transmission Electron Microscopy Cell for Crystal Structure Determination of Cathode Materials for Li-Ion batteries. *Nano Lett.* **2018**, *18*, 6286–6291. [CrossRef] [PubMed]

© 2019 by the author. Licensee MDPI, Basel, Switzerland. This article is an open access article distributed under the terms and conditions of the Creative Commons Attribution (CC BY) license (http://creativecommons.org/licenses/by/4.0/).

Article

In-Line Holography in Transmission Electron Microscopy for the Atomic Resolution Imaging of Single Particle of Radiation-Sensitive Matter

Elvio Carlino

Istituto per la Microelettronica ed i Microsistemi, Consiglio Nazionale delle Ricerche (CNR-IMM), Sezione di Lecce, Campus Universitario, via per Monteroni, 73100 Lecce, Italy; elvio.carlino@cnr.it

Received: 27 February 2020; Accepted: 18 March 2020; Published: 20 March 2020

Abstract: In this paper, for the first time it is shown how in-line holography in Transmission Electron Microscopy (TEM) enables the study of radiation-sensitive nanoparticles of organic and inorganic materials providing high-contrast holograms of single nanoparticles, while illuminating specimens with a density of current as low as 1–2 $e^-\text{Å}^{-2}\text{s}^{-1}$. This provides a powerful method for true single-particle atomic resolution imaging and opens up new perspectives for the study of soft matter in biology and materials science. The approach is not limited to a particular class of TEM specimens, such as homogenous samples or samples specially designed for a particular TEM experiment, but has better application in the study of those specimens with differences in shape, chemical composition, crystallography, and orientation, which cannot be currently addressed at atomic resolution.

Keywords: TEM; in-line holography; single particle imaging; atomic resolution imaging; radiation damage; soft matter; nanostructured drugs; organic materials

1. Introduction

Transmission Electron Microscopy (TEM) is widely used to study the properties of matter at the highest spatial resolution. There is a wide body of literature that reports on the study of single nanoparticles of inorganic material, showing how fundamental subtle physical effects can be understood by TEM experiments at atomic resolution [1]. High-Resolution TEM (HRTEM) enables direct access to the structural properties of individual particles, correlating their structure to their behavior. This allows the comprehension of matter at the atomic level and the development of new materials for a huge variety of applications [2]. In a realistic specimen, the particles are not necessarily all equal and properly oriented, but they could have different crystalline properties, defects, or allotropic state, which deeply influence the properties of the materials system. The study of single particles enables, in a batch of nanoparticles, the distinguishing of the differences between the particles and the relevant influence on the macroscopic behavior of the nanostructured material. In the case of radiation-sensitive material, standard HRTEM approaches on single particles could fail, due to the damage induced by the high-energy electrons on the specimen. This is a big issue for biologic materials [3]. Moreover, in pharmacy, TEM study of single nanocrystalline salt drugs, which are organic-like material consisting of low atomic number atoms tied with weak chemical bonds, cannot be performed by standard HRTEM [4,5]. In fact, radiation damage provides the main limitation to the spatial resolution of electron beam imaging or spectroscopy of organic materials [6]. The use of new TEM/STEM microscopes, equipped with aberration correctors and field emission cathodes, ascertains the highest spatial resolution so far achieved, but this technology can deliver a high-density of current on the specimen, making radiation damage an issue of growing importance also for inorganic materials and even metals [7]. As the radiation damage cannot be eliminated, there is a strong interest in finding approaches that can limit, or in some way overcome, the damage, enabling the study of the specimen at atomic resolution despite the radiation damage [8,9]. The issue of radiation damage is not only limited to electrons, but it is common to other probes, for example, X-rays. A very recent approach, which is polarizing the X-ray scientific community, is the so-called diffract-and-destroy method, developed for extremely intense sources of X-rays with high-frequency pulses, such as free electron lasers (FEL) [10]. In this case, the scattered photons are acquired on a femtosecond time-scale, and therefore before the explosion of the illuminated ensemble of molecules. Unfortunately, this approach cannot be straightforwardly extended to the case of electrons in a TEM, due to the peculiarity of electrons as charged particles and the relevant Boersch effect [9,11]. The radiation damage depends strongly on the specimen nature. Indeed, in a very schematic description, the knock-on damage is the main issue in conducting materials, whereas the ionization damage, and the related radiolysis, is the main issue in case of semiconducting or insulating inorganic materials and for all organic or organic-like matter. The knock-on damage is due to the deflection of the primary electrons by the electrostatic field of the nuclei of the specimen. For deflection angles up to 100 mrad, this scattering is considered elastic, to a good approximation, as the energy transferred to the specimen nucleus is below 0.1 eV. For angles of deflection higher than 100 mrad, this event can cause an energy transfer from the incident electron to the nuclei of the specimen of several eV, producing sputtering of the atoms from the specimen surface. For energy transfer of tens of eV, the atoms are displaced in the specimen, forming crystal defects [12]. In the case of ionization damage, known as radiolysis, the scattering considered is between electrons: the primary electrons lose part of their kinetic energy by ionizing the atoms of the specimen or exciting collective motion in the form of plasmons, which involves, in the case of metals, the vibration of atomic nuclei, but not their permanent displacements [6]. In the case of insulating or semi-insulating materials, the energy loss by the primary electrons produces holes that are not rapidly recombined, as they are in metals, but could result in a stable arrangement that stores the energy loss by the primary beam in a configuration of broken bonds [6]. Energy of few eV is enough to break a chemical bond, whereas the ionization energy involves tens of eV, and most of this energy is dissipated by producing secondary electrons. Consequently, an electron made available from an ionization process during its lifetime can break several bonds in an organic material, producing most of the damage in this class of materials [6]. The breaking of the bonds produces the loss of short-range order in crystalline materials and results in the appearance of a diffused halo in the diffraction pattern. It is a common and frustrating experience, during electron diffraction experiments in a TEM on radiation-sensitive materials that, as a function of the irradiation time and dose rate, the diffracted Bragg's spots are quickly progressively faded and disappear completely, as a result of the disruption of the crystalline order. An example is shown in Figure 1.

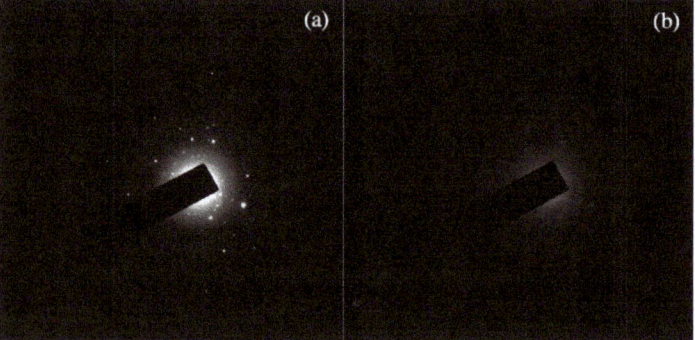

Figure 1. Selected area electron diffraction patterns of a specimen consisting of nanoparticles of Vincamine; (**a**) after an exposure to the electron beam with a density of current of 300 e$^-$Å^{-2}s^{-1} for 0.01 sec [4] and (**b**) after an exposure to the same current density for 0.3 s.

Figure 1 shows two selected area electron diffraction patterns obtained by illuminating an area of ~10 microns in diameter of a specimen consisting of nanoparticles of a salt of Vincamine [13], which is an indole alkaloid used for the treatment of important neurovegetative conditions, like Parkinson's and Alzheimer's diseases [14,15]. The current density of the electron probe is ~300 e$^-$Å^{-2}s^{-1}. In Figure 1a, the specimen has been exposed to the electrons for 0.01 sec, whereas in Figure 1b, the same area of the specimen is exposed to the electron beam for ~0.3 sec, which faded or cancelled the diffracted intensities, due to the damage of the crystalline structure. As reported by Egerton [6], the radiation damage is of particular concern in electron microscopy because of the need of high spatial resolution. Otherwise, we could simply defocus the incident beam and spread the damage over a large volume of the specimen; the fraction of the broken bonds would then become small and the radiation damage would cease to be a problem. It is therefore rather evident that imaging of single particle, in a specimen like the one shown in Figure 1, is quite cumbersome. Imaging can become impossible if we aim to look for a specific particle of interest before acquiring the atomic resolution HRTEM image. Indeed, the limit imposed by the radiation damage is particularly evident in the case of atomic resolution TEM study of nanoclusters, where when the beam is focused onto a beam sensitive particle, it can rapidly cause the disappearance of the crystalline order and, furthermore, the effect of charging could even produce a Coulomb explosion creating a hole in the eventual supporting film used for TEM observation. Therefore, especially in the case of a specimen consisting of nanoparticles, where the interest is to study the properties of single particles, before acquiring an atomic resolution HRTEM image, we need to find the particle suitable for the experiment. A suitable particle is defined as a particle properly oriented with respect to the electron beam to distinguish between eventual different crystal polytypes and orientation. The HRTEM image contrast is due mainly to the elastic scattering between the electrons and the nuclei of the atoms of the specimen [2]. Unfortunately, organic particles consist of low atomic number atoms, which have a relatively low elastic scattering power [2], and therefore a relatively high density of current is necessary to distinguish the particle with respect to the supporting film. Nevertheless, once the candidate particle is found, we need to evaluate the relevant diffraction pattern to check the proper orientation, therefore we need to wait until the eventual specimen holder drift is stopped and only at this point, after focusing, we can acquire the HRTEM image. Indeed, in the case of radiation-sensitive materials, we cannot even detect the particle by standard specimen survey conditions without destroying the particle itself or its eventual crystalline order. For example, in the case of radiation-sensitive polymers, to avoid quick degradation of the material, the electron current density threshold should be between 0.1 to 10 e$^-$Å^{-2}s^{-1} [16]. In these conditions, if we are looking for a particle in a TEM specimen with a low density of nanoparticles, we do not have enough image contrast to distinguish between the nanoparticle itself and the supporting carbon film. It is, therefore, important for a successful atomic resolution low-dose HRTEM experiment on radiation-sensitive specimen, to find an approach that enables the detection of an isolated particle, to check its crystallinity and orientation with respect to the electron beam, to check the specimen holder drift to enable atomic resolution imaging, and to adjust the electron optical conditions prior to low dose and low dose rate HRTEM acquisition, all by using an electron density of current between 0.1 to 10 e$^-$Å^{-2}s^{-1}. The use of expediencies to reduce the radiation damage in some classes of materials can be applied; this is the case, for example, in the use of a conducting coating on inorganic insulators to reduce the effect of radiolysis, the use of energy primary beam as low as 60 KeV to avoid the knock-on damage in many materials, or the use of low dose rate to enable the specimen recovery of knock-on damage in some classes of materials. There exists extensive literature on where these expediencies are considered and their effectiveness is discussed [6,9,17–20]. In organic materials and soft matter, a particular role is played by the use of low temperature to reduce the effect of radiolysis. In fact, for a long time, it has been recognized that the use of cryogenic temperature reduces the appearance of the effect of radiolysis in TEM images of proteins [21,22]. Nevertheless, the use of cryogenic temperature still needs the use of an extremely low dose of electrons, and this results in TEM images with extremely low contrast, where in some cases, the presence of a single particle in an electron micrograph can be hardly distinguished [21]. Actually, the effect of cryogenic temperature does not influence the ionization cross section, but reduces the desorption and the movement of the molecular fragments, whose bonds have been broken by the primary and secondary electrons. For example, at liquid nitrogen temperature, the density of electrons to image quite safely a protein has to be ≤5 e^- $Å^{-2}$, whereas at the liquid helium temperature, it has to be ≤ 20 e^- $Å^{-2}$ [21]. This is why nowadays the approach, which is revolutionizing the structural biology in TEM, is the cryo-EM, which is a method that enables the study of proteins that cannot be easily crystallized for study by X-ray crystallography, for example, the membrane proteins. The word "cryo" denotes that this method is performed at cryogenic temperature. The development of this approach earned the authors the 2017 Nobel Prize in Chemistry [23]. Cryo-EM was actually established as method decades ago and requires the acquisition of thousands of low dose images from a specimen consisting of identical particles. The images are then processed by sophisticated software that produces a tridimensional model of the particle. Note that cryo-EM is not a true imaging approach, and the resulting structural model needs to be validated by well-controlled protocols, whose development is still in progress with dedicated task forces. The validation of structural models is a field where there is a strong research activity, and the number of structures solved by cryo-EM is growing fast [24]. The biggest success in the last years of cryo-EM is due to some particular technological advances that represent a turning point in the results achievable by cryo-EM, producing the so-called "Resolution Revolution" [25]. One key technological advance for the practical use of cryo-EM has been the development of direct conversion detectors to acquire the electron images with much better performances for the same low dose of electrons [26,27]. The other key point has been the development of capacity of calculus, which was hard to conceive when the basics of cryo-EM were proposed. These aspects were underlined by Richard Henderson during his Nobel Prize acceptance speech. It is worthwhile to remark once again, that the result of a so-called single-particle imaging by cryo-EM, is not a true single particle imaging at all, as it uses the images of thousands of particles, assumed as identical, to produce a 3-D model of the macromolecule. This reconstruction is, in any case, achieved at a spatial resolution worse than the one allowed by the electron optics, and related to the statistics of the particles imaged and also to the accuracy of the starting model of the particle to be imaged.

Here, we report on the use of in-line holography in TEM to perform true single-particle atomic resolution imaging of soft matter and biologic nanoparticles, believed not accessible by high-energy atomic resolution TEM experiments. In the reported studies, the experiments were successfully performed at room temperature using 200 KeV electrons. In-line holography is used to survey the specimen, to find suitable isolated particles, and to tune the experimental conditions to enable a reliable quantitative atomic resolution imaging experiment on radiation-sensitive materials. This approach enables to acquire safely low dose and low dose rate HRTEM images from radiation-sensitive materials, gathering information in analogy with the well-established methodology used in materials science on specimens robust to electron irradiation.

2. Methods

The imaging and analysis performances of a TEM, reported by the manufactures, have meaning only if the numbers of electrons "N", measured within each spatial resolution element, have a statistical significance. The radiation damage limits the number of electrons for each resolution element of size δ and, therefore, is directly correlated to the resolution [28,29]. If D_c is the critical electron dose that the specimen region δ can tolerate without damage then

$$N = F(\frac{D_c}{e})\delta^2 \tag{1}$$

where "e" is the electron charge and F is the fraction of electrons reaching the detector. The image contrast C, between the recorded pixel and its neighbors containing N_b electrons, is defined as

$$C = (N - N_b)/N_b = \Delta N/N_b \qquad (2)$$

(C is negative in the case of absorption contrast).

The signal to noise ratio (SNR) is:

$$(SNR) = \sqrt{(DQE)}(\Delta N/N_{shot}) \qquad (3)$$

where DQE is the detective quantum efficiency, which is a measure of the noise introduced by the detector, and $N_{shot} = \sqrt{N + N_b}$, according to Poisson's statistics.

Therefore, from Equations (1)–(3), the size of δ depends linearly on the SNR. The dose-limited resolution is thus

$$\delta = \frac{(SNR)}{\sqrt{(DQE)}} \frac{\sqrt{2}}{|C|} \frac{1}{\sqrt{FD_c/e}} \qquad (4)$$

Equation (4) gives analytical evidence, in the approximation of weak contrast [30], of the role of the radiation damage on the resolution. It is worthwhile to mark the role of F, which makes bright-field imaging intrinsically less efficient than phase contrast in terms of radiation damage limited resolution [29], irrespective to the peculiar features of the two imaging methods. Moreover, the role of DQE should be noted, as it is the reason why, in the last years after the introduction of direct conversion detectors with better DQE, we are observing a fast growth of microscopes equipped with direct conversion devices especially on the instruments dedicated to cryo-EM, but not limited to them. The Rose's criterion states that, to distinguish two adjacent elements in an image, the SNR has to be at least 5 [31]. From Equation (4) and considering $F = 1$, for HRTEM, and a perfect detector, resulting in DQE = 1

$$\delta = \left(5\frac{\sqrt{2}}{|C|}\right)\sqrt{\frac{e}{D_c}} \qquad (5)$$

According to this equation, we can estimate the resolution limit for a polymer with a reasonable D_c of 0.01 C/cm². Considering a contrast of 20% (C = 0.2) it results δ~1 nm [32]. The high sensitivity of some classes of materials, like polymers, nanodrugs, or biologic matter, limits the resolution in an imaging experiment, due to the need to use low dose of electrons. Furthermore, these materials have a relatively small cross section for elastic scattering with electrons resulting in a poor image contrast. This makes it impossible to perform an experiment of atomic resolution imaging on single radiation-sensitive nanoparticle, as its intrinsic low scattering power, together with the fast damage, limits the particle visibility and the tuning of the electron optical conditions for quantitative imaging before its damage. An example is shown in Figure 1. Indeed, by using standard survey methods, like bright-field or HRTEM, to find the particles suitable for quantitative atomic resolution imaging experiments, we are groping in the dark looking for where the representative nanoparticles are. Highly defocused electron probe coupled to strong defocus of the objective lens, or eventual future dedicated phase plates with improved stability with respect to charging effect and contaminations [33] could help to increase the contrast in the sample survey to detect the nanoparticles. Nevertheless, once the particle has been seen, we still do not have any clue of its crystallinity and orientation, and in the time necessary to put in focus the objective lens and to stabilize the eventual drift for an atomic resolution image, the particle would have been destroyed or, at least, it would be no longer representative of the pristine particle. The use of an in-line hologram could overcome these difficulties as it provides high contrast evidence of the nanoparticle, even if it is composed of light chemical elements, while providing clues on its crystalline status and on the electron optical conditions. All this required using density of electron current of <1 e⁻Å⁻²s⁻¹. In-line electron holography was proposed in 1948 by Dennis Gabor for overcoming the limitation related to electron lens aberrations [34], and his idea was awarded in 1971 by the Nobel prize in physics. The in-line hologram formation is schematized in Figure 2.

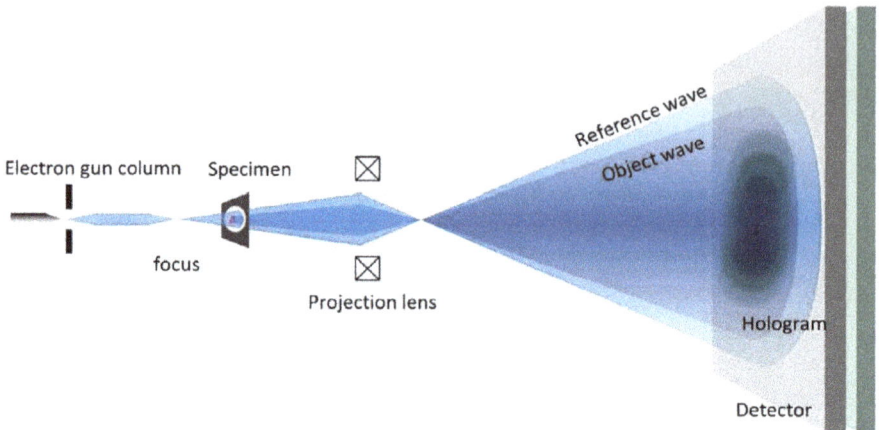

Figure 2. Scheme of the in-line hologram as firstly proposed by Dennis Gabor.

Electron holography in TEM historically followed, for its application in physics, biology, and materials science, mostly another experimental configuration called off-axis holography, which requires a reference wave field obtained by splitting the illuminating wave field before the interaction with the specimen by using, in most of the cases, an electron biprism [35]. The reference is then used to recover the wave field after the scattering with the specimen. The introduction of the off-axis holography is related to practical reasons for quantitative applications of the holography to overcome the "twin image problem" related to the original proposal of Gabor [36–39]. Although for many years after its proposal in-line holography was not used for quantitative imaging, this experimental configuration is well known, for example, to all those electron microscopists involved in convergent beam electron diffraction (CBED) or scanning transmission electron microscopy (STEM). Indeed, this approach was used, for example, as a visual aid in CBED to follow the movement of the specimen in the direct space (shadow image) while observing the diffraction pattern in the reciprocal space during tilting for accurate specimen orientation with respect to the electron beam [40]; or in STEM, to accurately tune the microscope illumination lenses alignment necessary for accurate quantitative experiments. In-line holograms are also used for lens aberrations measurements and corrections [41], for accurate focusing [42], and for a variety of applications, as recognized in the pioneering work of J. M. Cowley [43]. In fact, as far as the detection of a low scattering nanoparticle concerns, the presence of the twin-image effect is an advantage as it enhances the contrast of the particle in the hologram. This feature of the in-line hologram, is here used to set up the method for the very low dose survey and to tune the electron optical set up to enable HRTEM image on single particle. It is indeed worthwhile to mark here that, recently, thanks to the advances in digital holography, and in particular in phase-shifting digital holography [44], the "twin image problem" has been successfully overcome, enabling an in-line hologram reconstruction free from twin-image disturbance, and therefore making feasible the retrieval of the relevant intensity and phase distribution [45], establishing in-line holography itself as a possible quantitative approach to atomic resolution imaging.

The experimental conditions in the in-line holography, and in particular the electron current density in the area of interest, can be readily changed by simply acting on the microscope illumination conditions, and therefore in-line holograms can be formed and observed in the reciprocal space varying the density of the electrons on the specimen to extremely low value, but still effective to detect the shadow image of the particles and, at the same time, looking at diffraction coming from the illuminated area, as shown in Figure 3. When the electron probe is focused above or below the specimen plane, each diffracted disc in the reciprocal plane contains a shadow image of the direct plane of the specimen. The magnification "M" of this shadow image is related to distance "u" between the focal plane and the specimen, and to the distance "v" between the specimen plane and the plane of view. From the geometric optics, M = v/u. The magnification of the image can be readily changed by changing the plane where the electron probe is focused [46].

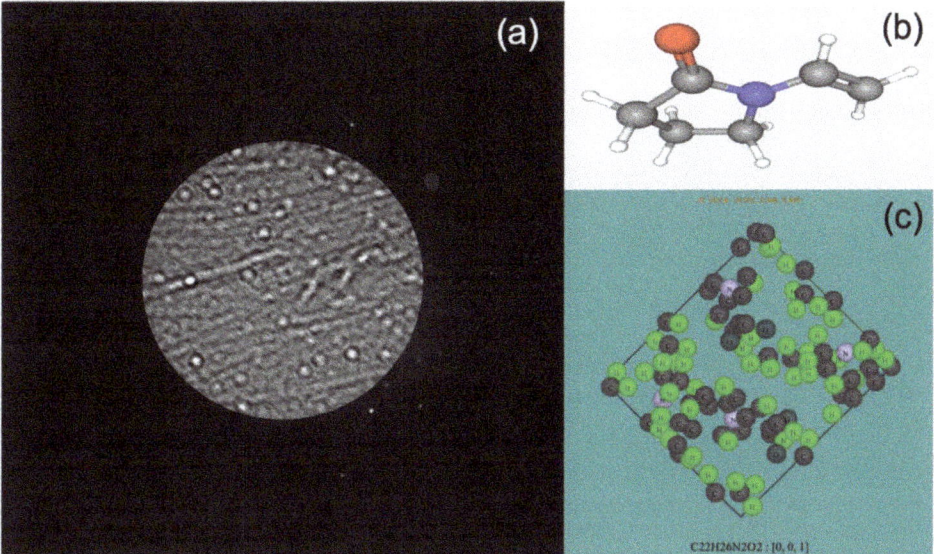

Figure 3. (a) in line-hologram on Vinpocetine and polyvinylpyrrolidone acquired in diffraction mode. (b) 3D Chemical structure of polyvinylpyrrolidone (gray atoms: C; white atoms: H; red atoms: O; blue atoms: N). (c) Crystal cell of Vinpocetine in [0,0,1] zone axis.

The hologram in Figure 3 comes from an area of ~150 nm in diameter, illuminated by an electron density of ~0.1 e^-Å^{-2}. The specimen consists of vinpocetine and polyvinylpyrrolidone (PVP), and was studied at room temperature as no cooling holder was necessary, despite the extreme sensitivity of the material to the radiation damage. The electron energy was 200 KeV. The low dose of electrons enables surveying of the sample finding an area suitable for atomic resolution imaging with no detectable degradation of the particles. The size of the circular area in the diffractogram is related to the size of the used condenser aperture, whereas the magnification of the particles and their contrast are related to the illumination condition and in particular to the spatial coherence of the probe [47]. In the circular area in the central part of Figure 3, the presence of spherical particles of vinpocetine and filaments of polyvinylpyrrolidone is clearly detectable in the shadow image, despite the low intensity of the electron probe and the low atomic number of the atoms in both vinpocetine and PVP [48,49]. The shadow image also enables the accurate monitoring of the eventual drift of the specimen and checking when the conditions are such that an HRTEM from the particle of interest can be safely recorded with a low dose rate. In the dark area around the region selected by the condenser aperture, the sharp diffracted intensities are the signature of the crystalline nature of some of the structures in the illuminated area. The advantage of operating in the reciprocal space is evident, as the hologram contains all the information on the shape and crystal status of the particle of interest and also on the electron optical conditions, whereas the focus of the objective lens, adjusted in the direct space on an area close to the area of interest before switching the electron optics to diffraction mode, remains fixed and it is recovered immediately by switching back the electron optics to conjugate the direct space to the detector. This is the optimal condition to acquiring the relevant low dose and low dose rate HRTEM image, as shown in Figure 4.

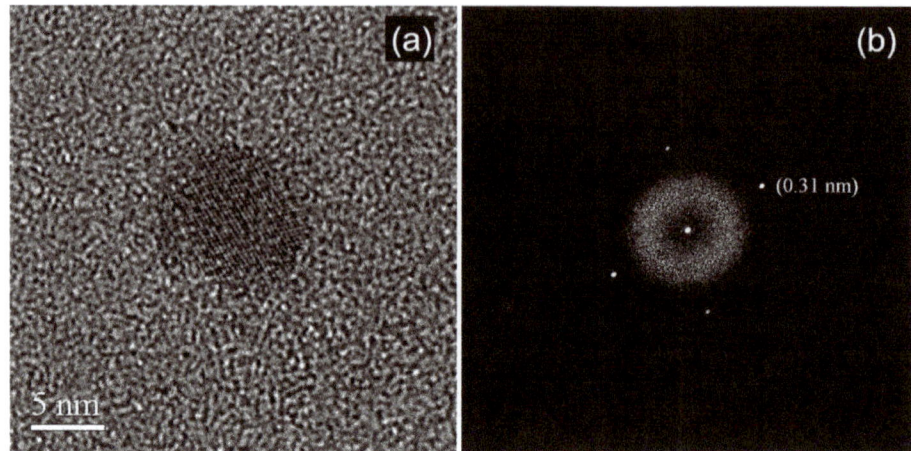

Figure 4. Low dose rate HRTEM image (**a**) and relevant diffractogram from a crystalline particle of Vinpocetine (**b**). The lattice spacing measured on the diffractogram is reported.

Indeed, in Figure 4, the HRTEM image of a particle of vinpocetine together with the relevant diffractogram is shown. The density of electrons to image the particle is ~100 e$^-$ Å^{-2}. The lattice contrast in the particle is rather sharp indicating a high degree of crystal order, as also confirmed by the relevant diffractogram. Note that the experiment was performed at room temperature by using a JEOL 2010 FEG UHR TEM/STEM (Jeol ltd., Tokyo, Japan) operated at 200 kV. The FEG cathode enables to illuminate the specimen by a highly coherent probe of electrons, whereas the Cs = (0.47 ± 0.01) mm of the objective lens provides an interpretable spatial resolution at optimum defocus for HRTEM of 0.19 nm [2]. The equipment was operated in free lens control to finely tune the illumination conditions [4,5]. The beam current was measured by Faraday's cup. This equipment and the above reported experimental conditions were used for a variety of successful experiments on radiation-sensitive materials, and some of these experiments are discussed in the next paragraph.

3. Results and Discussion

The method for atomic resolution imaging of radiation-sensitive materials by in-line holography coupled to HRTEM has been extensively applied, in our laboratory, to drug salts, but also to biologic samples, enabling the achieving of atomic resolution imaging despite the use of primary electrons of 200 KeV and specimen at room temperature [4,5]. The results shown in this paragraph focus on the low dose and low dose rate HRTEM experiments, giving significant atomic resolution insights of the nanoparticles of radiation-sensitive matter, however note that the HRTEM experiments follow the in-line holography survey of the specimen and the procedures, described in the methods section, necessary for accurate survey, electron optical tuning, and low-dose HRTEM experiments at low dose rate.

in the following, the results on vincamine nanoparticles (Section 3.1), co-crystals nanoparticles of caffeine and glutaric acid (Section 3.2), and nanoparticles of ferrihydrite bound to creatinine particles (Section 3.3) are reported.

3.1. Vincamine

Here, the in-line holography method was applied to the study of the nanoparticles of vincamine citrate as obtained by "solid-excipient assisted mechanochemical salification". The aim was to correlate the enhancement of solubilization kinetics of ball-milled vincamine citrate with respect to the vincamine citrate obtained by classical synthetic routes and, more generally, to understand the structural origin of the different features of this drug obtained by different methods of synthesis [4]. The analysis of single particles repeated on hundreds of particles, within the limit of a statistical analysis by HRTEM experiments, enables acquiring information on the different crystallographic properties related to the differences in the material preparation. Moreover, this leads to a better understanding of the results of the X-ray diffraction pattern measurements, as far as the peak broadening is concerned, and photoelectron spectroscopy experiments performed on the same specimens [4]. The HRTEM images in Figures 5 and 6, with the relevant diffractograms, were obtained by exposing the particles to a parallel electron beam and illuminating a relatively large area, of ~100 nm in diameter, of the specimen around the particle of interest. The density of electrons in the illuminated areas was of ~100 $e^- Å^{-2}$ and no evidence of electron-induced damage was detected. It is worthwhile to remark that during these kinds of experiments, we noticed the role of the dose rate on the particle damage, namely, low dose rate has relatively little effect on the particles damage [50], at the same total dose delivered, as the structure has the time to recover the damage between different collisions events and in images series at low dose rate the particles can maintain their pristine structure [51,52]. During the in-line holography survey with low density of electron current, typically between 0.1 and 10 $e^- Å^{-2} s^{-1}$, the features of the diffracted spots remained unchanged for tens of seconds but, if the density of electron current is increased, by changing the excitation of the condenser lens in a range of 10^2 to 10^3 $e^- Å^{-2} s^{-1}$, the diffracted spots become immediately faded and disappear. Figure 5 shows a particle of vincamine oriented along a high symmetry zone axis, in the relevant diffractogram the spots correspond to a lattice spacing of (0.20 ± 0.01) nm, which is a typical value for many polytypes of vincamine. The experiments show that about 10% of the observed particles have a crystalline nature, whereas in the remaining cases, the particles are amorphous. In the case of the crystalline particles, the availability of measurable symmetries in the diffractogram enables not only to measure the lattice spacing but also to compare the experimental symmetries with those simulated by using the known allotropic states of Vincamine. This results in a higher accuracy in the relative measurement of spacing in the same diffractogram. As a result, the comparison between the experimental diffractograms and the simulated ones indicates a deformation of the crystal cell possibly due to the synthesis process in the presence of solid excipients [4].

The use of low dose rate during the acquisition of the HRTEM images enables to detect and study the extended structural defects in the pristine nanoparticles. This was done, for example, as shown in Figure 6, where a stacking fault, approximately in the middle of the nanoparticle, is detectable in the lattice fringes contrast, and it is reflected in the splitting of the relevant spots in the diffractogram on the right of the figure. In the diffractogram, the split concerns two couples of intensities corresponding to a spacing of (0.30 ± 0.01) nm, whereas the remaining couple of diffracted beams are due to a spacing of (0.32 ± 0.01) nm. The presence of extended defects in the structure of crystalline vincamine is related to the mechanochemistry process used for its synthesis, and it is at the origin of the broadening of the peaks in the relevant XRD measurements [4].

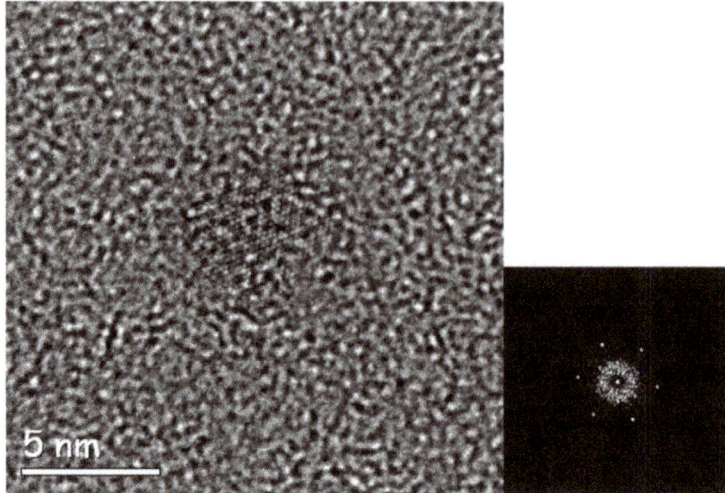

Figure 5. Left: Low dose rate HRTEM of a particle of vincamine. Right: the relevant high symmetry diffractogram (reprinted by courtesy from Hasa et al. [4]).

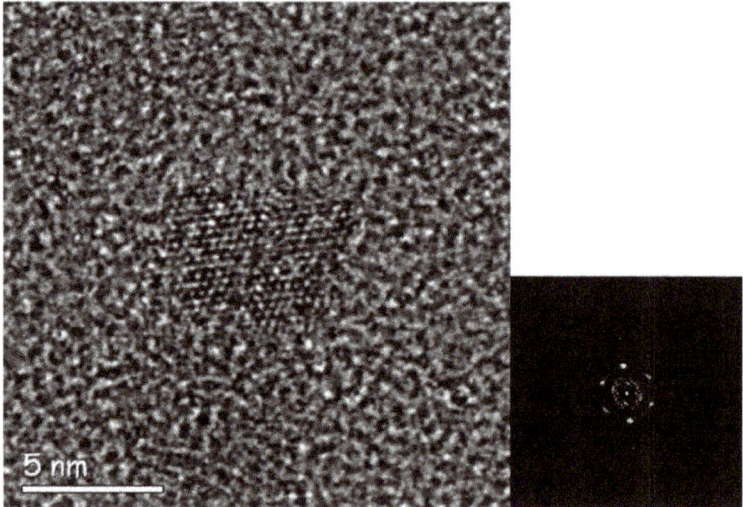

Figure 6. Left: Low dose rate HRTEM of a particle of vincamine. Right: the relevant high symmetry diffractogram. The lattice fringes in the central part of the particle and the spot splitting in the diffractogram point the presence of an extended structural defect in the crystal structure (reprinted by courtesy from Hasa et al. [4]).

3.2. Caffeine/Glutaric Acid Co-Crystals

The in-line holography-based approach was applied here to the study of co-crystallization in nanoparticles, used in the design of a supramolecular structure with desired functional properties. Indeed, a co-crystal is a solid having two or three different molecules in the crystal structure and, therefore, it is particularly attractive for the application in engineering of composition of pharmaceutical phases [53]. For example, a molecule active against a particular disease can be associated in the same crystal cell of another drug, which is capable to reach a particular target or it is capable to overcome the cell barriers. on the other hand, co-crystals can easily exhibit polymorphism that can have deep influence on the properties of a drug, as demonstrated by the case of the anti-HIV drug ritonavir [54]. Single-particle studies by TEM in-line holography-based atomic resolution imaging have the possibility to access the crystal properties of individual nanoparticle of the drug, revealing the polytype and the influence of the synthesis process in controlling polymorphism phenomena [5]. The mechanochemical co-crystallization reaction in polymer-assisted grinding represents a well-controlled approach to the co-crystallization process, but it needs an appropriate understanding of the influence of polymer structure and polarity, together with the grinding conditions, on the synthesis results and, therefore, on the polytypes synthetized [55]. Caffeine (CAF) and glutaric acid (GLA) represent an ideal case study for the understanding of the co-crystallization process [5]; in particular, as solid excipient, ethylene glycol polymer chains of variable length and polarity were used. An in-line hologram of caffeine-glutaric acid (CAF-GLA) co-crystals is shown in Figure 7. TEM specimens were prepared dispersing the pristine powders on a copper grid previously covered by an amorphous carbon film, avoiding any pre-dispersion in liquid to prevent their possible modification. The aim is to have a low density of pristine nanoparticles on the copper grid to avoid accidental modification of the particle structure in the area illuminated by the electron beam, but not in the field of view of the microscopist and not under his direct monitoring.

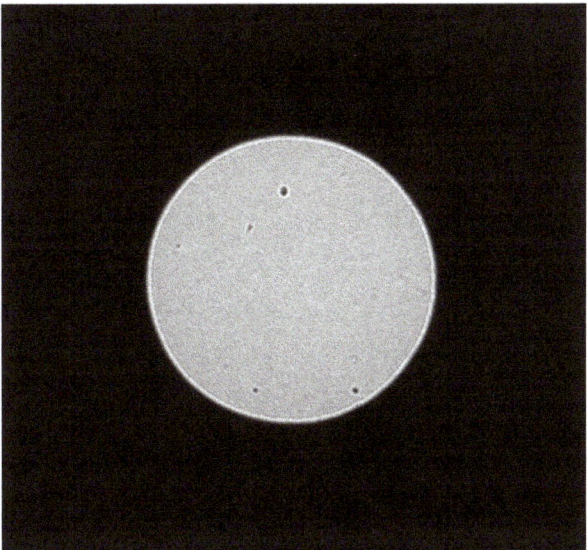

Figure 7. In line-hologram on cocrystals of caffeine and glutaric acid as acquired in diffraction mode and exposed to 1.2 e$^-$ Å^{-2}.

The in-line hologram in Figure 7 is acquired in diffraction mode and shows few nanoparticles of ~10 nm in diameter, with high contrast in a field of view of ~500 nm. The density of electrons in the in-line hologram is ~1.2 e$^-$Å^{-2}. The low density of particles, on one hand, reduces the possibility of artifacts but, on the other hand, requires frequent and relatively wide movements of the specimen holder, with relevant specimen drift, to locate the particles and to put them properly in the field of view. It is therefore important to have an extremely low-dose approach, like the in-line hologram with defocused illumination, to check also the specimen drift until it stops, before switching the electron optical conditions to an intrinsically higher dose mode in the direct space HRTEM imaging, for an appropriate low dose exposure time. Figure 8 summarizes the atomic resolution information gained from one particle.

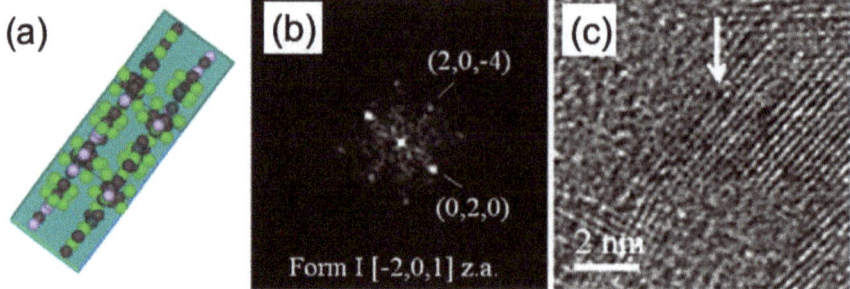

Figure 8. Caffeine-glutaric acid (CAF-GLA) co-crystal of polytype I [5]. The kind of polytype of the nanoparticle is univocally determined by comparing the structure of type I, viewed along the [−2, 0, 1] zone axis in (**a**), with the experimental diffractogram (**b**) and the HRTEM image (**c**). The arrow points the region from which the diffractogram was extracted (reprinted by courtesy from Hasa et al. [5]).

The particle in Figure 8 was synthetized by using a chain of polyethylene glycol of 1000 monomers as solid excipient. Figure 8c shows the HRTEM zoom on a nanoparticle oriented along a high symmetry zone axis, whose diffractogram in the area pointed by the arrow is shown in Figure 8b. The identification of the polytype can be univocally performed by comparing the experimental results with the simulations performed by using the Crystallographic Information File (CIF) available for the different polytypes of CAF-GLA system in the crystallography open database. The simulations here were performed by JEMS [56]. In particular, the particle in Figure 8 is a polytype I CAF-GLA oriented along the [−2, 0, 1] zone axis. The same approach was applied to the particle in Figure 9.

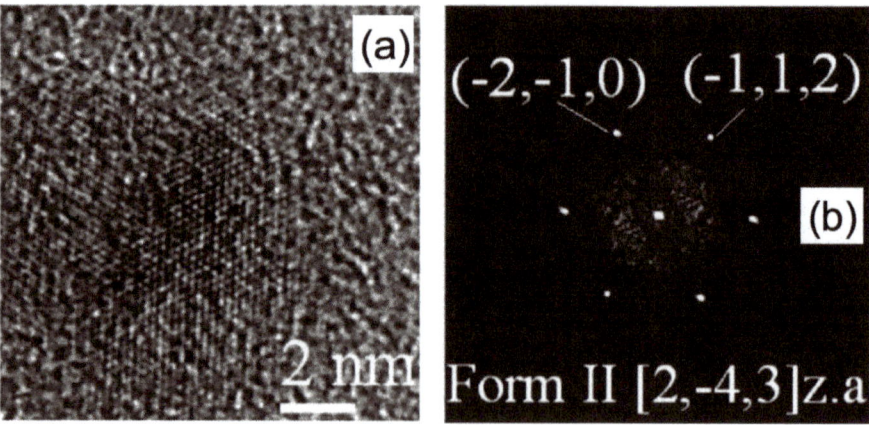

Figure 9. (**a**) HRTEM image of a caffeine (CAF) glutaric acid (GLA) polytype II [5] particle along with (**b**) the relevant diffractogram (reprinted by courtesy from Hasa et al. [5]).

In the latter case, the particle belongs to the polytype II of CAF-GLA system and the relevant diffractogram shows that the particle is oriented along the [2, −4, 3] zone axis with respect to the primary electron beam. These kinds of experiments enable the study of the crystallography and the morphologic properties of individual pristine particles addressing the role of the synthesis conditions on the structure and the properties of the CAF-GLA co-crystals [5]. This approach enables the application of well-known and powerful electron microscopy methods, developed in materials science for materials robust to the radiation damage, to radiation-sensitive single particles.

3.3. Creatinine-Ferrihydrite Nanoparticles

In this last example of applications, the in-line holography-based atomic resolution imaging approach was applied to biologic traces of creatinine bound to ferrihydrate, which can be present in the bloodstream of patients suffering from acute kidney disease [57]. The pristine particles were placed on a copper grid and inserted in the high vacuum of the TEM specimen chamber without any pretreatment, like staining or similar procedures, usually employed on biologic specimens to increase the image contrast, or coating with carbon or metals, to prevent the charging effect and to partially protect the specimen from the electron irradiation. The experiments were performed at room temperature at an acceleration voltage of 200 kV. Figure 10a shows the in-line hologram as acquired, in diffraction mode, from a group of particles. The dark part on the left of Figure 10a is due to the mesh of the copper grid. The density of electron current is ~0.5 $e^-Å^{-2}s^{-1}$, and the illuminated area is ~3.5 micron in diameter. Note that most of the particles visible in the in-line hologram show the evidence of some substructures. The origin of these substructures can be immediately recognized from the low magnification HRTEM image in Figure 10b, acquired on one of the particles visible in Figure 10a, where dark small particles appear embedded within the big one. The morphology, the size, and the contrast of the particle in Figure 10b are similar to those of globular protein of ferritin [58], but a further investigation rules this interpretation out.

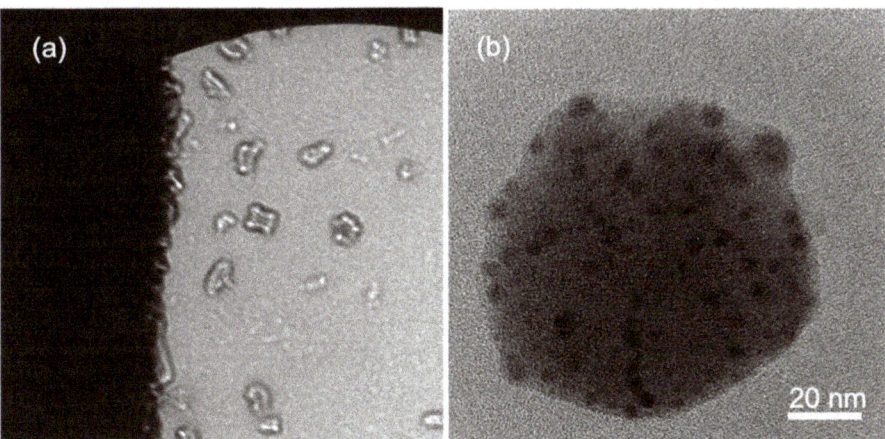

Figure 10. (a) In-line hologram of biologic particles. (b) Low-magnification HRTEM of one of the biologic nanoparticles; note the dark smaller nanoparticles within the big one.

As matter of the fact, the structure of the small dark particles, as measured from all the diffractograms, is compatible with the ferrihydrite of the ferritin, as shown, as an example, in Figure 11. In this case, the experimental data and the relevant simulation, performed by considering the known crystal structure of ferrihydrite [59], enable to index the particle as ferrihydrate oriented along the [4, 2, 1] zone axis with respect to the primary electron beam. In the figure are also reported the relevant Miller's indexes together with their spacing. Nevertheless, all of the experimental data, HRTEM images, and relevant diffractograms acquired in the big particles away from the dark particles never reproduce what is simulated by using the known crystal structure of the globular protein of ferritin. Furthermore, we caution the reader that in all of our experiments we never observed the ferrihydrite fingerprint spacing at 0.25 nm. The discrepancy in the experimental data with respect to the hypothesis suggested by the morphologies and some structural data of the ferrihydrite can be understood in the light of what was reported in some studies of the interaction between ferrihydrite nanoparticles and creatinine and urea [56]. Note that the interaction and bonding between ferrihydrite and creatinine, or urea, is not likely to happen in the blood of a healthy organism, but could occur when some pathologic events, for example, rhabdomyolysis, determine the occurrence of the interaction of the content of the muscle cells, like ferritin, with waste substances, like urea and creatinine, contained in the blood stream. Indeed, rhabdomyolysis is a serious syndrome due to a direct or indirect muscle injury, and it results from the death of muscle fibers and release of their contents into the bloodstream. This can lead to serious complications, such as fatal renal failure, as the kidneys cannot remove waste and concentrated urine [60].

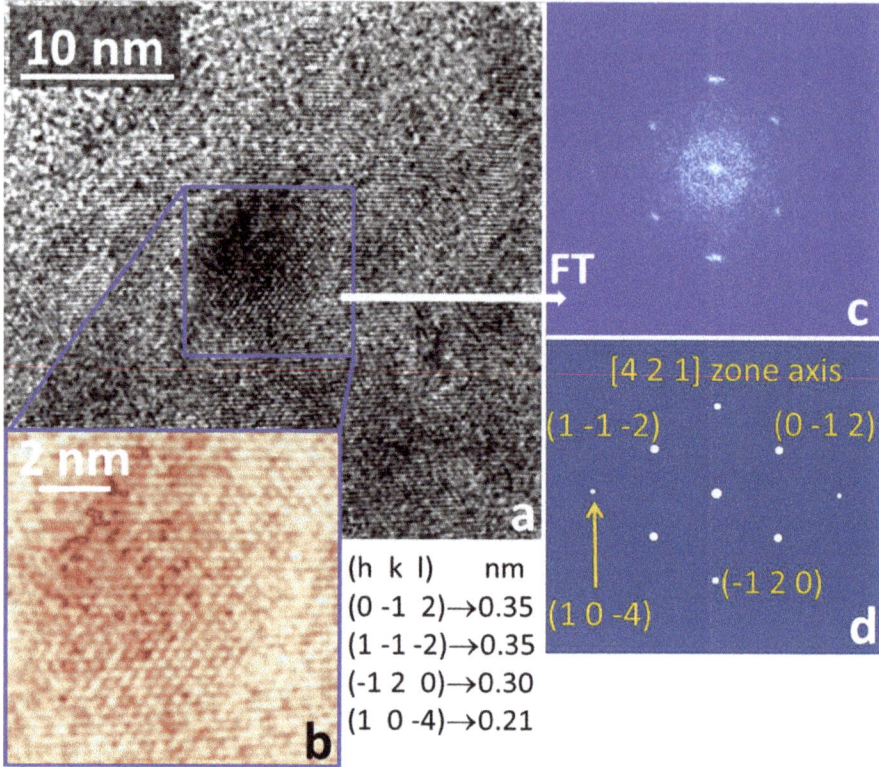

Figure 11. Representative results on the small dark particles. (**a**) HRTEM image focused on one of the small and dark nanoparticles shown in Figure 10; (**b**) zoom inside the blue square of panel (**a**) (false color output); (**c**) Fourier transform relevant to the zoomed area in panel (**b**); (**d**) simulation, with the Miller indexes associated to some spots and the corresponding lattice spacing. Simulations show that the pattern of panel (**c**) is compatible with the ferrihydrate in the [4, 2, 1] zone axis.

The studies by X-ray diffraction pattern reported in literature on the interactions between iron oxide nanoparticles and creatinine and urea show that the iron oxide fingerprint spacing at 0.25 nm is always absent when the nanoparticles of iron oxide are bound to creatinine or urea [56].

This evidence suggested the comparison of the experimental TEM results of the crystalline structure of the nanoparticle, like the one shown in Figure 10b, away from the iron oxide dark particles (Figure 11a,b), with the simulation performed starting from the known structure of creatinine and urea. The results of these comparisons show that in all the HRTEM experimental data collected, the relevant Fourier transforms are compatible with the structure of creatinine. An example is shown in Figure 12. Figure 12a is a HRTEM image focused on an area of the big envelope particle. Note the darker contrast due to the ferrihydrite nanoparticles and the lighter contrast due to the structure of the big particle embedding the ferrihydrite nanoparticles. Figure 12b shows the lattice contrast of the HRTEM image zoomed in the region marked by the blue square in Figure 12a. The relevant Fourier transform is shown in Figure 12c. The latter is compared with the pattern simulated starting from the known crystallographic information file (CIF) of creatinine structure shown in Figure 12d. In the latter, there are also reported the relevant Miller's indexes of the crystal planes together with the spacing.

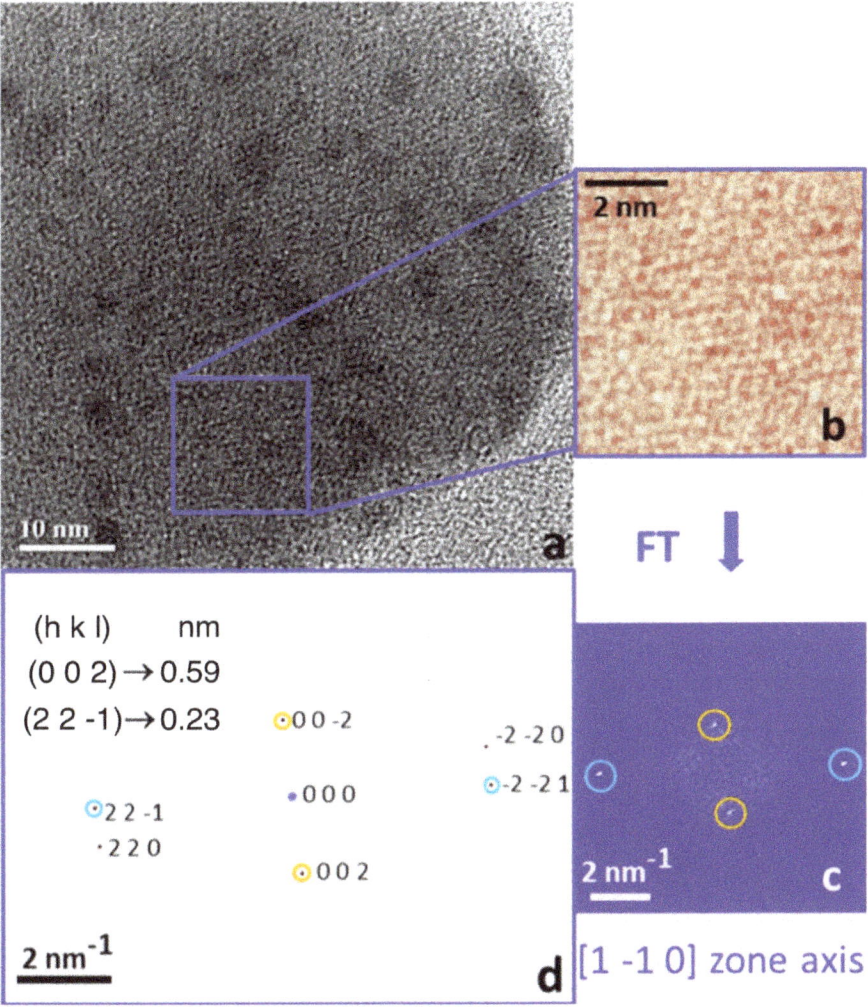

Figure 12. (a) Characterization of creatinine particles; HRTEM image; (b) magnified view of the square region marked in (a); (c) Fourier transform of (b); (d): simulation of the diffraction pattern of creatinine in [1, −1, 0] zone axis with reported the lattice spacing relevant to the observed intensities. The yellow and pale-blue circles in the diffraction pattern simulation mark the correspondence with the circles around the experimental spots in panel (c).

As a result of the TEM experiments and simulations, the particles present in the analyzed biological traces are therefore due to creatinine bound to ferrihydrate nanoparticles. This result helps understand that the biological traces studied by TEM could be due to the serum of an organism with a pathologic bonding of creatinine to ferrihydrite due to rhabdomyolysis. The detailed experimental data on the biologic specimen, collected at room temperature and by using the pristine material, were successfully obtained due to the capability of the in-line holography low dose approach to study a specimen by high-energy electrons despite its sensitivity to the radiation damage. The approach achieves high spatial resolution information of individual biologic particles as whole, and within their inner structure, enabling to detect eventual anisotropy within their volume.

4. Conclusions and Future Perspectives

Atomic resolution single-particle TEM studies of radiation-sensitive organic and inorganic matter are essential for the complete understanding of the properties of this important class of materials and therefore for the advances in biology, pharmacology, medicine, material science, physics, etc. Accurate true imaging of single particles is also a prerequisite for a successful and more reliable structural modeling by cryo-EM, providing a priori information that drives the modeling. Unfortunately, the sensitivity to radiation damage prevents a straightforward use of the powerful TEM/STEM atomic resolution methods developed for materials robust to radiation as not only the particle irradiation for the imaging acquisition itself, but all the steps necessary for a meaningful quantitative single particle imaging experiment can destroy, or at least damage, the case of interest.

Here, it was established how in-line holography coupled with HRTEM enables the performing of extremely low dose experiments on radiation-sensitive nanoparticles of organic and inorganic materials, thus accessing the properties of single particles allowing the understanding of their structural properties and enabling the correlation to their performances. All of these experiments were not on specially designed specimens, but on standard specimens of pristine nanoparticles with different structure, chemistry, and morphology. This true single-particle study of radiation-sensitive matter opens new perspective in a variety of scientific disciplines. In particular, here the use of in-line holography allows setting up the electron optics to provide reliable low dose and low dose rate atomic resolution imaging, to find the particle of interest accessing immediately its crystalline status and shape, to monitor the eventual specimen drift, and to check when it stops to make atomic resolution imaging possible. All these steps delivering a safe density of current between 0.1 to 10 $e^- \text{Å}^{-2} s^{-1}$ and monitoring the eventual structural damage. The experiments on nanoparticles of drugs and of organic materials have shown that the pristine properties of single particles of radiation-sensitive matter can be studied at atomic resolution even at room temperature by using electrons of 200 keV.

The results shown so far regard the combination of in-line holography and low dose rate HRTEM to provide true atomic resolution imaging on single particle. Moreover, the recent understanding on in-line holography and digital reconstruction of in-line holograms [44,45,61] indicates a further advancement in the application of in-line holography to provide, by itself, detailed information of the atomic structure of organic molecules while delivering low dose of electrons to the specimen. These features together with the successful theoretical demonstration that tridimensional structure of organic molecules can be recovered by in-line holography from a single projection [52] pave the way to a new direct knowledge of the atomic structure of organic and inorganic materials by electrons in a TEM.

Funding: This work was partially supported by the NOXSS PRIN (2012Z3N9R9) project and Progetto premiale MIUR 2013 USCEF.

Acknowledgments: I would like to acknowledge L. De Caro for the extensive fruitful interactions to interpret the experimental data of the creatinine–ferrihydrate system and for the help in realizing Figures 11 and 12. I would like also to thank A. Taurino for the careful reading of the paper.

Conflicts of Interest: The authors declare no conflicts of interest.

References

1. Van Aert, S.; Batenburg, K.J.; Rossell, M.D.; Erni, R.; Van Tandeloo, G. Three-dimensional atomic imaging of crystalline nanoparticles. *Nature* **2011**, *470*, 374–377. [CrossRef] [PubMed]
2. Spence, J.C.H. *Experimental High-Resolution Electron Microscopy*, 2nd ed.; Oxford University Press Inc.: New York, NY, USA, 1988; ISBN 0-19-505405-9.
3. Wisedchaisri, G.; Reichow, S.L.; Gonen, T. Advances in structural and functional analysis of membrane proteins by electron crystallography. *Structure* **2011**, *19*, 1381–1393. [CrossRef] [PubMed]
4. Hasa, D.; Perissutti, B.; Cepek, C.; Bhardwaj, S.; Carlino, E.; Grassi, M.; Invernizzi, S.; Voinovich, D. Drug salt formation via mechanochemistry: The case study of vincamine. *Mol. Pharm.* **2013**, *10*, 211–224. [CrossRef] [PubMed]
5. Hasa, D.; Carlino, E.; Jones, W. Polymer-assisted grinding, a versatile method for polymorph control of cocrystallization. *Cryst. Growth Des.* **2016**, *16*, 1772–1779. [CrossRef]
6. Egerton, R.F. Control of radiation damage in the TEM. *Ultramicroscopy* **2013**, *127*, 100–108. [CrossRef]
7. Malac, M.; Beleggia, M.; Egerton, R.; Zhu, Y. Cs corrected bright field TEM imaging of radiation sensitive materials. *Microsc. Microanal.* **2005**, *2*, 2150–2151. [CrossRef]
8. Malac, M.; Beleggia, M.; Taniguchi, Y.; Egerton, R.F.; Zhu, Y. Low-dose performance of parallel-beam nanodiffraction. *Ultramicroscopy* **2008**, *109*, 14–21. [CrossRef]
9. Egerton, R.F. Outrun radiation damage with electrons? *Adv. Struct. Chem. Imaging* **2015**, *1*, 1–11. [CrossRef]
10. Chapman, H.N.; Fromme, P.; Barty, A.; White, T.A.; Kirian, R.A.; Aquila, A.; Hunter, M.S.; Schulz, J.; DePonte, D.P.; Weierstall, U.; et al. Femtosecond X-ray protein nanocrystallography. *Nature* **2011**, *470*, 73–77. [CrossRef]
11. Knauer, W. Boersch effect in electron-optical instruments. *J. Vac. Sci. Technol.* **1979**, *16*, 1676–1679. [CrossRef]
12. Egerton, R.F.; McLeod, R.; Wang, F.; Malac, M. Basic questions related to electron-induced sputtering in the TEM. *Ultramicroscopy* **2010**, *110*, 991–997. [CrossRef]
13. Vereczkey, L. Pharmacokinetics and metabolism of vincamine and related compounds. *Eur. J. Drug Metab. Pharmacokinet.* **1985**, *10*, 89–103. [CrossRef]
14. Karpati, E.; Biro, K.; Kukorelli, T. Investigation of vasoactive agents with indole skeletons at Richter Ltd. *Acta Pharm. Hung.* **2002**, *72*, 25–36.
15. Vas, A.; Gulyas, B. Eburnamine derivatives and the brain. *Med. Res. Rev.* **2005**, *25*, 737–757. [CrossRef]
16. Leijten, Z.J.W.A.; Keizer, A.D.A.; de With, G.; Friedrich, H. Quantitative analysis of electron Beam damage in organic thin films. *J. Phys. Chem. C* **2017**, *121*, 10552–10561. [CrossRef]
17. Egerton, R.F.; Li, P.; Malac, M. Radiation damage in the TEM and SEM. *Micron* **2004**, *35*, 399–409. [CrossRef]
18. Hobbs, L.W. Radiation effects in analysis by TEM. In *Introduction to Analytical Electron Microscopy*; Hren, J.J., Goldstein, J.I., Joy, D.C., Eds.; Plenum Press: New York, NY, USA, 1987; pp. 399–445. ISBN 978-1-4757-5581-7.
19. Nan, J.; Spence, J.C.H. on the dose-rate threshold of beam damage in TEM. *Ultramicroscopy* **2012**, *113*, 77–82.
20. Meyer, J.C.; Kotakoski, J.; Mangler, C. Atomic structure from large area, low dose exposure of materials: A new route to circumvent radiation damage. *Ultramicroscopy* **2014**, *145*, 13–21. [CrossRef]
21. Henderson, R. Cryo-protection of protein crystals against radiation damage in electron and X-ray diffractions. *Proc. R. Soc. Lond.* **1990**, *B241*, 6–8.
22. Unwin, P.N.T.; Henderson, R. Molecular structure determination by electron microscopy of unstained crystalline specimens. *J. Mol. Biol.* **1975**, *94*, 425–440. [CrossRef]
23. Available online: https://www.nobelprize.org/prizes/chemistry/2017/press-release/ (accessed on 14 February 2020).
24. Henderson, R.; Sali, A.; Baker, M.L.; Carragher, B.; Devkota, B.; Downing, K.H.; Egelman, E.H.; Feng, Z.; Frank, J.; Grigorieff, N.; et al. Outcome of the first electron microscopy validation task force meeting. *Structure* **2012**, *20*, 205–214. [CrossRef] [PubMed]
25. Kühlbrandt, W. The resolution revolution. *Science* **2014**, *343*, 1443–1444. [CrossRef] [PubMed]
26. Faruqi, A.R.; Henderson, R. Electronic detectors for electron microscopy. *Curr. Opin. Struct. Biol.* **2007**, *5*, 549–555. [CrossRef]
27. Merk, A.; Bartesaghi, A.; Banerjee, A.; Falconieri, V.; Rao, P.; Davis, M.I.; Pragani, R.; Boxer, M.B.; Earl, L.A.; Milne, J.L.S.; et al. Breaking Cryo-EM resolution barriers to facilitate drug discovery. *Cell* **2016**, *165*, 1698–1707. [CrossRef] [PubMed]
28. Egerton, R.F. Mechanisms of radiation damage in beam-sensitive specimens, for TEM accelerating voltages between 10 and 300 kV. *Microsc. Res. Tech.* **2012**, *75*, 1550–1556. [CrossRef] [PubMed]

29. Egerton, R.F. Choice of operating voltage for a transmission electron microscope. *Ultramicroscopy* **2014**, *145*, 85–93. [CrossRef] [PubMed]
30. Egerton, R.F. Radiation damage to organic and inorganic specimens in the TEM. *Micron* **2019**, *119*, 72–87. [CrossRef] [PubMed]
31. Rose, A. Television pickup tubes and the problem of vision. In *Advances in Electronics*; Marton, L., Ed.; Academic Press: New York, NY, USA, 1948; pp. 131–166.
32. Libera, M.L.; Egerton, R.F. Advances in the transmission electron microscopy of polymers. *Polym. Rev.* **2010**, *50*, 321–339. [CrossRef]
33. Van Dyck, D. Wave reconstruction in TEM using a variable phase plate. *Ultramicroscopy* **2010**, *110*, 571–572. [CrossRef]
34. Gabor, D. A new microscopic principle. *Nature* **1948**, *161*, 777–778. [CrossRef]
35. Lichte, H. Electron holography approaching atomic resolution. *Ultramicroscopy* **1986**, *20*, 293–304. [CrossRef]
36. Lohmann, A. Optische Einseitenbandübertragung angewandt auf das Gabor-Mikroskop. *Opt. Acta* **1956**, *3*, 97. [CrossRef]
37. Leith, E.; Upatnieks, J. Reconstructed wavefronts and communication theory. *JOSA* **1962**, *52*, 1123–1128. [CrossRef]
38. Leith, E.N.; Upatnieks, J. Wavefront reconstruction with continuous-tone objects. *JOSA* **1963**, *53*, 1377. [CrossRef]
39. Testorf, M.; Lohmann, A.W. Holography in phase space. *Appl. Opt.* **2008**, *47*, A70–A77. [CrossRef]
40. Steeds, J.W.; Carlino, E. *Electron crystallography*. In *Electron Microscopy in Materials Science*; Merli, P.G., Vittori-Antisari, M., Eds.; World Scientific Publishing Co. Pte. Ltd.: Singapore; Hackensack, NJ, USA; London, UK; Hong Kong, 1992; pp. 279–313. ISBN 981-02-0924-X.
41. Lupini, A.R.; Wang, P.; Nellist, P.D.; Kirkland, A.I.; Pennycook, S.J. Aberration measurement using the Ronchigram contrast transfer function. *Ultramicroscopy* **2010**, *110*, 891–898. [CrossRef]
42. Nellist, P.D. Scanning Transmission Electron Microscopy. In *Science of Microscopy*; Hawkes, P., Spence, J.C.H., Eds.; Springer: New York, NY, USA, 2007; Volume 1, pp. 65–132. ISBN 978-0-387-24296-4. [CrossRef]
43. Cowley, J.M. Electron diffraction phenomena observed with a high-resolution STEM instrument. *Microsc. Res. Tech.* **1986**, *3*, 25–44. [CrossRef]
44. Goodman, J.W. *Introduction to Fourier Optics*, 4th ed.; Freeman W. H. and Co. Macmillan Learning: New York, NY, USA, 2017; p. 427, ISBN-13 978-1-319-11916-4.
45. Latychevskaia, T.; Fink, H.W. Solution to the twin image problem in holography. *Phys. Rev. Lett.* **2007**, *98*, 233901–233904. [CrossRef]
46. Spence, J.C.H.; Carpenter, R.W. Electron microdiffraction. In *Principles of Analytical Electron Microscopy*; Joy, D.C., Romig, A.D., Jr., Goldstein, J., Eds.; Plenum Press: New York, NY, USA; London, UK, 1986; p. 326.
47. Ronchi, V. Forty years of history of a grating interferometer. *Appl. Opt.* **1964**, *3*, 437–450. [CrossRef]
48. Available online: https://pubchem.ncbi.nlm.nih.gov/compound/Vinpocetine (accessed on 14 February 2020).
49. Available online: https://pubchem.ncbi.nlm.nih.gov/compound/n-vinyl-2-pyrrolidone (accessed on 14 February 2020).
50. Chen, J.Z.A.; Sachse, C.; Xu, C.; Mielke, T.; Spahn, C.M.T.; Grigorieff, N. A dose rate effect in Single-particle electron microscopy. *J. Struct. Biol.* **2008**, *161*, 92–100. [CrossRef]
51. Bloemen, M.; Brullot, W.; Thien Luong, T.; Geukens, N.; Gils, A.; Verbiest, T. Improved functionalization of oleic acid-coated iron oxide nanoparticles for biomedical applications. *J. Nanopart. Res.* **2012**, *14*, 1100. [CrossRef] [PubMed]
52. Chen, F.-R.; Kisielowski, C.; Van Dyck, D. Prospects for atomic resolution in-line holography for a 3D determination of atomic structures from single projections. *Adv. Struct. Chem. Imaging* **2017**, *3*, 1–7. [CrossRef] [PubMed]
53. Fleischman, S.G.; Kuduva, S.S.; McMahon, J.A.; Moulton, B.; Bailey Walsh, R.D.; Rodriguez-Hornedo, N.; Zaworotko, M.J. Crystal Engineering of the composition of pharmaceutical phases: Multiplecomponent crystalline solids involving Carbamazepine. *Cryst. Growth Des.* **2003**, *3*, 909–919. [CrossRef]
54. Bauer, J.; Spanton, S.; Henry, R.; Quick, J.; Dziki, W.; Porter, W.; Morris, J. Ritonavir: An extraordinary example of conformational polymorphism. *Pharm. Res.* **2001**, *18*, 859–866. [CrossRef] [PubMed]
55. Hasa, D.; Schneider Rauber, G.; Voinovich, D.; Jones, W. Cocrystal formation through mechanochemistry: From neat and liquid-assisted grinding to polymer-assisted grinding. *Angew. Chem. Int. Ed.* **2015**, *54*, 7371–7375. [CrossRef] [PubMed]
56. J E M S-S a A S. Version 4.3931U2016. Available online: http://www.jems-saas.ch/ (accessed on 14 February 2020).
57. Banerji, B.; Pramanik, S.K. Binding studies of creatinine and urea on iron nanoparticle. *Springerplus* **2015**, *4*, 708. [CrossRef]

58. Quintana, C.; Cowley, J.M.; Marhic, C. Electron nanodiffraction and high-resolution electron microscopy studies of the structure and composition of physiological and pathological ferritin. *J. Struct. Biol.* **2004**, *147*, 166–178. [CrossRef]
59. Janney, D.E.; Cowley, J.M.; Buseck, P.R. Structure of synthetic 2-line ferrihydrite by electron nanodiffraction. *Am. Mineral.* **2000**, *85*, 1180–1187. [CrossRef]
60. Torres, P.A.; Helmstetter, J.A.; Kaye, M.A.; Kaye, A.D. Rhabdomyolysis: Pathogenesis, diagnosis, and treatment. *Ochsner J.* **2015**, *15*, 58–69.
61. Latychevskaia, T.; Fink, H.W. Practical algorithms for the simulation and reconstruction of digital in-line holograms. *Appl. Opt.* **2015**, *54*, 2424–2434. [CrossRef]

© 2020 by the author. Licensee MDPI, Basel, Switzerland. This article is an open access article distributed under the terms and conditions of the Creative Commons Attribution (CC BY) license (http://creativecommons.org/licenses/by/4.0/).

Review

Holography and Coherent Diffraction Imaging with Low-(30–250 eV) and High-(80–300 keV) Energy Electrons: History, Principles, and Recent Trends

Tatiana Latychevskaia [1,2]

[1] Physics Institute, University of Zurich, Winterthurerstrasse 190, 8057 Zurich, Switzerland; tatiana@physik.uzh.ch
[2] Paul Scherrer Institute, Forschungsstrasse 111, 5232 Villigen, Switzerland

Received: 17 June 2020; Accepted: 7 July 2020; Published: 10 July 2020

Abstract: In this paper, we present the theoretical background to electron scattering in an atomic potential and the differences between low- and high-energy electrons interacting with matter. We discuss several interferometric techniques that can be realized with low- and high-energy electrons and which can be applied to the imaging of non-crystalline samples and individual macromolecules, including in-line holography, point projection microscopy, off-axis holography, and coherent diffraction imaging. The advantages of using low- and high-energy electrons for particular experiments are examined, and experimental schemes for holography and coherent diffraction imaging are compared.

Keywords: holography; electron holography; in-line holography; diffraction; coherent diffraction imaging; iterative phase retrieval; biomolecules

---

1. Introduction

We present the theoretical background to electron scattering in an atomic potential, and highlight the differences between low- and high-energy electrons interacting with matter. This theoretical introduction provides the background and definitions necessary for the remaining sections. We then present several interferometric techniques that can be realized with electrons, following the chronological order in which they were discovered (in-line holography, point projection microscopy, off-axis holography, and coherent diffraction imaging), and provide some examples. The advantages and disadvantages of various techniques realized with low- and high-energy electrons are discussed.

2. Electron Waves

2.1. The Wavelength of an Electron

In 1924, Louis de Broglie published his PhD thesis entitled "Research on the Theory of the Quanta" [1], in which he described the hypothesis that with every particle of matter with mass m and velocity v a real wave must be 'associated' and defined the wavelength

$$\lambda = \frac{h}{p}, \tag{1}$$

where h is the Planck's constant and p is the momentum. Although de Broglie formulated his hypothesis for any type of particle, he won the Nobel Prize for Physics in 1929 for his discovery of the wave nature of electrons, after their wave-like behavior was first experimentally demonstrated in 1927 by Clinton Davisson and Lester Germer [2].

The wavelength of an accelerated electron in an electron microscope can be derived from the energy-momentum relation

$$E_{total}^2 = (pc)^2 + (m_0 c^2)^2 \qquad (2)$$

and the total energy of the electron

$$E_{total} = m_0 c^2 + eU. \qquad (3)$$

Here, m_0 is the electron's rest mass and eU is the kinetic energy of the electron, which arises from the accelerating voltage U. By combining Equations (2) and (3), we obtain an expression for the momentum, p

$$p = \frac{1}{c}\sqrt{eU(2m_0 c^2 + eU)}. \qquad (4)$$

By substituting p from Equation (4) into Equation (1) we obtain the wavelength of the electron

$$\lambda = \frac{hc}{\sqrt{eU(2m_0 c^2 + eU)}}. \qquad (5)$$

The values of this wavelength can range from 8.59 pm for a 20 keV electron to 1.97 pm for a 300 keV electron, where 20–300 keV is the range of energies for medium energy transmission electron microscope (TEM), which are the most widespread equipment nowadays for the study of organic and inorganic matter.

For low-energy electrons with kinetic energies of up to 300 eV, the relativistic effects can be neglected, and the wavelength of the electron can be calculated as

$$\lambda = \frac{h}{\sqrt{2m_0 eU}}. \qquad (6)$$

Here, the values of the wavelength range from 1.73 Å for a 50 eV electron to 0.78 Å for 250 eV.

2.2. Electron Scattering in the First Born Approximation

2.2.1. The Schrödinger Equation

Electrons are scattered by atomic potentials, and the wavefunction of a particle moving in a potential $V(\vec{r})$ is described by the Schrödinger equation

$$\left[-\frac{\hbar^2}{2m}\nabla^2 + V(\vec{r})\right]\psi(\vec{r}) = E\psi(\vec{r}), \qquad (7)$$

where $\psi(\vec{r})$ is the wavefunction of the electron, E is its energy, and $\hbar$ is the reduced Planck's constant $\hbar = h/(2\pi)$. The Schrödinger equation can be re-arranged as follows:

$$(\nabla^2 + k^2)\psi(\vec{r}) = \frac{2m}{\hbar^2}V(\vec{r})\psi(\vec{r}), \qquad (8)$$

where we have introduced $k^2 = \frac{2m}{\hbar^2}E$. The solution to the Schrödinger equation (Equation (8)) can be written in the form:

$$\psi(\vec{r}) = \psi_0(\vec{r}) + \frac{2m}{\hbar^2}\iiint G(\vec{r}_0 - \vec{r})V(\vec{r}_0)\psi(\vec{r}_0)d\vec{r}_0, \qquad (9)$$

where $\psi_0(\vec{r})$ is the solution to the homogeneous equation $(\nabla^2 + k^2)\psi_0(\vec{r}) = 0$ and $G(\vec{r})$ is the solution to $(\nabla^2 + k^2)G(\vec{r}) = \delta(\vec{r})$. $G(\vec{r})$ is the so-called Green function: $G^{\pm}(\vec{r}) = -\frac{1}{4\pi}\frac{e^{\pm ikr}}{r}$, which describes convergent (−) or divergent (+) spherical waves. For a stationary scattering wave, we can choose $G(\vec{r}) = G^+(\vec{r})$, and can rewrite the solution to the Schrödinger equation Equation (9) as follows:

$$\psi(\vec{r}) = \psi_0(\vec{r}) - \frac{m}{2\pi\hbar^2} \iiint \frac{e^{ik|\vec{r}-\vec{r}_0|}}{|\vec{r}-\vec{r}_0|} V(\vec{r}_0)\psi(\vec{r}_0) d\vec{r}_0. \tag{10}$$

Next, we will make use of the fact that the scattering potential $V(\vec{r})$ is localized within a small region, meaning that $r_0 \approx 0$. Since we are interested in the electron wavefunction far from the scattering center, we can use the approximation $r \gg r_0$ and can then expand

$$|\vec{r} - \vec{r}_0| = \sqrt{r^2 - 2\vec{r}\vec{r}_0 + r_0^2} \approx r\sqrt{1 - \frac{2\vec{r}\vec{r}_0}{r^2}} \approx r - \frac{\vec{r}\vec{r}_0}{r}. \tag{11}$$

In Equation (11) and everywhere else in the text, scalar product of vectors is used, unless otherwise specified. Using the approximation in Equation (11), we rewrite Equation (10) as

$$\psi(\vec{r}) = \psi_0(\vec{r}) - \frac{m}{2\pi\hbar^2} \frac{e^{ikr}}{r} \iiint e^{-i\vec{k}\vec{r}_0} V(\vec{r}_0)\psi(\vec{r}_0) d\vec{r}_0 \tag{12}$$

where we have introduced $\vec{k} = k\frac{\vec{r}}{r}$ as the wave vector of the scattered wave.

2.2.2. The Born Approximation

Note that in Equation (12), the solution for $\psi(\vec{r})$ contains the same function $\psi(\vec{r}_0)$ in the integral. We must therefore search for a solution by applying a series of approximations. We apply the Born approximation, which originates from perturbation theory and considers only the first term of the series expansion. In the zero-th order approximation, we keep only the first term: $\psi(\vec{r}) \approx \psi_0(\vec{r})$. In the first-order approximation, we use the $\psi(\vec{r})$ found in the zero-th order approximation and substitute it into the integral given in Equation (12):

$$\psi_1(\vec{r}) \approx \psi_0(\vec{r}) - \frac{m}{2\pi\hbar^2} \frac{e^{ikr}}{r} \iiint e^{-i\vec{k}\vec{r}_0} V(\vec{r}_0)\psi_0(\vec{r}_0) d\vec{r}_0. \tag{13}$$

In the second-order approximation, we use the $\psi_1(\vec{r})$ found in the first order approximation (Equation (13)) and substitute it into the integral given by Equation (12):

$$\psi_2(\vec{r}) \approx \psi_0(\vec{r}) - \frac{m}{2\pi\hbar^2} \frac{e^{ikr}}{r} \iiint e^{-i\vec{k}\vec{r}_0} V(\vec{r}_0)\psi_1(\vec{r}_0) d\vec{r}_0, \tag{14}$$

and so forth. Typically, the first-order approximation is sufficient to describe the scattered wavefront (Equation (13)).

2.2.3. Scattering Amplitude

For a plane incident wave $\psi_0(\vec{r}) = Ae^{ikz}$, the scattered wave in the first-order Born approximation is given by:

$$\psi(\vec{r}) = Ae^{ikz} - \frac{m}{2\pi\hbar^2} \frac{e^{ikr}}{r} \iiint e^{-i\vec{k}\vec{r}_0} V(\vec{r}_0)\psi_0(\vec{r}_0) d\vec{r}_0, \tag{15}$$

which can be re-written as

$$\psi(\vec{r}) = Ae^{ikz} + f(\vartheta,\varphi) A \frac{e^{ikr}}{r}. \tag{16}$$

Here, Ae^{ikz} is the incident plane wave, $A\frac{e^{ikr}}{r}$ is the outgoing spherical wave (as also described by the Huygens-Fresnel principle), and $f(\vartheta,\varphi)$ is the complex-valued scattering amplitude:

$$f(\vartheta,\varphi) = -\frac{m}{2\pi\hbar^2 A}\iiint e^{-i\vec{k}\vec{r}_0}V(\vec{r}_0)\psi_0(\vec{r}_0)\mathrm{d}\vec{r}_0. \quad (17)$$

The incident plane wave can be rewritten as $\psi_0(\vec{r}_0) = Ae^{ikz_0} = Ae^{i\vec{k}_0\vec{r}_0}$, where $\vec{k}_0 = k\vec{e}_z$ is the wave vector of the incident plane wave. This geometrical arrangement and the symbols used are illustrated in Figure 1. By re-writing Equation (17), we get the result that, in the first-order Born approximation, the scattering amplitude is the Fourier transform (FT) of the scattering potential:

$$f(\vartheta,\varphi) = -\frac{m}{2\pi\hbar^2 A}\iiint e^{-i(\vec{k}-\vec{k}_0)\vec{r}_0}V(\vec{r}_0)\mathrm{d}\vec{r}_0. \quad (18)$$

The differential scattering cross-section is given through the scattering amplitude as:

$$\frac{\mathrm{d}\sigma}{\mathrm{d}\Omega} = |f(\vartheta,\varphi)|^2. \quad (19)$$

In Equation (19), σ is the elastic cross section, where the elastic scattering events form the signal of most of the imaging methods in TEM and also in both holographic and coherent diffraction imaging methods.

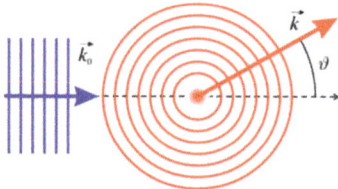

Figure 1. Schematic of the electron scattering event and illustration of the symbols used. $\vec{k}_0$ and $\vec{k}$ are the wave vectors of the incident plane wave and the scattered wave, respectively, and ϑ is the scattering angle.

2.2.4. Examples of Scattering Amplitudes

Differential scattering cross-sections $\mathrm{d}\sigma/\mathrm{d}\Omega$, calculated as a function of the scattering angle for carbon (C) and gold (Au) atoms, are shown in Figure 2 for electrons with energy 150 and 200 keV. The differential scattering cross-sections were calculated using the NIST Electron Elastic-Scattering Cross-Section Database (version 3.2, National Institute of Standards and Technology, Gaithersburg, MD, USA) [3]. From the plots shown in Figure 2, we can draw the following conclusions. The scattering amplitude of the electrons is maximal in the direction of the incident beam. Low-energy electrons (150 eV) scatter with a maximal amplitude within a cone of 40–60°, while high-energy electrons (200 keV) scatter with maximal amplitude within a very narrow cone of up to 1°. Elements with higher atomic numbers scatter more strongly; for example, 200 keV electrons are scattered 15 times more strongly by Au (atomic number 79) than by C (atomic number 6).

2.2.5. Inelastic Mean Free Path (IMFP) for High- and Low-Energy Electrons

An important difference between low- and high-energy electrons is the inelastic mean free path (IMFP), which is the average distance between inelastic scattering events. IMFP defines the maximal thickness of the samples which can be imaged with electrons. When an electron beam is propagating through a material, it loses intensity according to the expression

$$I(z) = I_0\exp\left(-\frac{z}{\lambda_i}\right) \quad (20)$$

where λ_i is the IMFP. A generalized expression for the IMFP as a function of electron energy was derived by Seah and Dench as follows [4]:

$$\lambda_i = \frac{143}{E^2} + 0.054\sqrt{E}, \tag{21}$$

where E is the electron energy in eV and λ_i is the IMFP in nm, as plotted in Figure 3a. The IMFP for low-energy electrons, measured experimentally using carbon films, is shown in Figure 3b. For low-energy electrons, the IMFP is on the order of a few Angstroms, which implies that only samples that are a few Angstroms thick can be measured in transmission mode. For high-energy electrons (with typical electron energies of 80–300 keV in TEMs), the IMFP ranges from tens to hundreds of nanometers, allowing us to probe thicker samples in transmission mode.

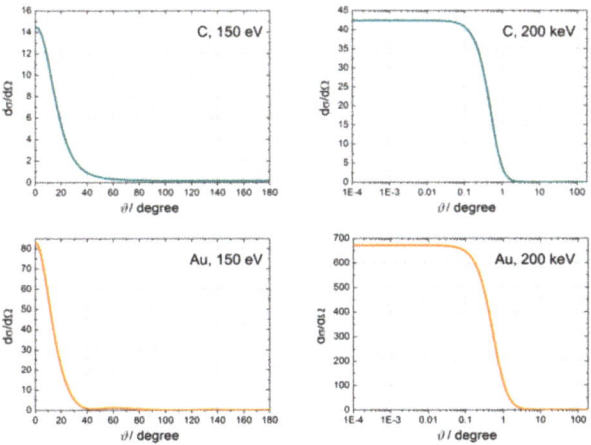

Figure 2. Differential scattering cross-sections $d\sigma/d\Omega$ of 150 eV and 200 keV electrons scattered by C and Au atoms, calculated as a function of the scattering angle ϑ. The units for the differential cross-sections are a_0^2/sr, where a_0 is the Bohr radius.

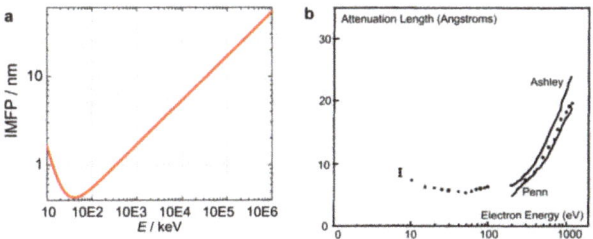

Figure 3. Inelastic mean free path (IMFP) as a function of electron energy. (**a**) IMFP calculated according to Equation (21). (**b**) IMFP measured based on the transmission of an electron beam through a thin amorphous carbon film as a function of the kinetic energy of the electron beam, using a transmission energy loss spectrometer [5]. Continuous lines are theoretical predictions for the IMFP by Ashley [6] and Penn [7], respectively; reprinted from [5], with permission from Elsevier.

2.3. Transmission Function, Object Phase, Exit Wave, and Phase Problem

In physics, and particularly in optics, the term "phase" is often used for a range of different phenomena. To clarify this terminology, we consider the typical arrangement of an optical experiment, as shown in Figure 4. An incident wavefront propagates through a sample and then toward a detector, where the intensity is measured.

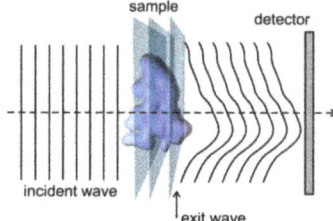

Figure 4. Schematic illustrating the object phase, exit wave, and phase problem.

Each plane in the sample can be represented by a transmission function:

$$t(x,y) = \exp[-a(x,y)]\exp[i\varphi(x,y)], \qquad (22)$$

where the distribution $a(x,y)$ describes the absorbing properties of the sample and the distribution $\varphi(x,y)$ describes the phase shift introduced by the sample into the probing wave. $\varphi(x,y)$ is called the object phase. After the incident wave propagates through the entire sample, the complex-valued distribution of the wavefront immediately behind the sample is called the exit wave. The distribution of this exit wave is often reconstructed from an experimental record. Finally, the exit wave propagates toward a detector, where the distribution of the wavefront can be written as $U(X,Y) = U_0(X,Y)e^{i\Phi(X,Y)}$, where $U_0(X,Y)$ is the amplitude and $\Phi(X,Y)$ is the phase distribution. Since a detector can only record the intensity $I(X,Y) = |U_0(X,Y)|^2$, information about the phase distribution is completely lost. However, this information is important in reconstructing the complete complex-valued wavefront in the detector plane, since the phase distribution contains information about the individual scattering events that took place inside the sample. Hence, to reconstruct the sample, we not only need to record the intensity of the diffracted wave, but must also know or determine its phase. This constitutes the phase problem.

2.4. Phase Shift of an Electron Wave in Electric and Magnetic Fields

When an electron is moving in an electric or magnetic potential, its wavefunction gains an additional phase shift compared to that of an electron moving through a region without a potential. In this section, we derive the phase shifts for an electron moving in both electric and magnetic potentials.

2.4.1. Phase Shift of an Electron Wave in an Electric Potential

The wavefunction of an accelerated electron moving in an electric potential is described by the time-dependent Schrödinger equation:

$$\left[-\frac{\hbar^2}{2m}\nabla^2 + V(\vec{r})\right]\Psi(\vec{r},t) = i\hbar\frac{\partial \Psi(\vec{r},t)}{\partial t}, \qquad (23)$$

where the time-dependent component of the eigenfunction $\Psi(\vec{r},t)$ is given by $\exp(-iEt/\hbar)$. For a particle with charge q, its energy E depends on the electrostatic potential, and in a region with a constant potential V, the potential energy qV is added to E, resulting in an additional phase shift of

$$\Delta\varphi = -\frac{qV}{\hbar}t, \qquad (24)$$

where t is the time spent in the potential. For a region with potential $V(x,y,z)$, the phase shift of an electron $q = -e$ moving along the z-direction is given by:

$$\Delta\varphi_E = \frac{e}{\hbar v} \int_{\text{path}} V(x,y,z)\mathrm{d}z = \frac{e}{\hbar v} V_z(x,y) = \sigma V_z(x,y), \qquad (25)$$

where v is the velocity of the electron, and $V_z(x,y)$ is the potential $V(\vec{r})$ projected on the (x,y) plane:

$$V_z(x,y) = \int V(x,y,z)\mathrm{d}z. \qquad (26)$$

In Equation (25), we introduced the interaction parameter

$$\sigma = \frac{e}{\hbar v}. \qquad (27)$$

2.4.2. Transmission Functions

Figure 5a shows the projected potentials for C and Au atoms, which were calculated using the parameterized atomic potentials, as explained in reference [8]. The results in Figure 5a show that the Au atom has a much stronger projected potential than the C atom, and as a result introduces a much stronger phase shift. The phase shifts for the 150 eV and 200 keV electrons used to probe the C and Au atoms, calculated by Equation (25), are shown in Figure 5b,c, respectively. The phase shift depends on the projected potential and the interaction parameter. The interaction parameter is larger for low-energy electrons and it is relatively small for high-energy electrons [9]. For electrons of energy 200 keV, the estimated phase shift at $r = 0.1$ Å is 1.04 rad for Au and 0.16 rad for C. Thus, when imaged with 200 keV electrons, the C atom can be considered a weak phase object. The transmission function of a weak phase object can be approximated as

$$t(x,y) = \exp[i\sigma V_z(x,y)] \approx 1 + \sigma V_z(x,y). \qquad (28)$$

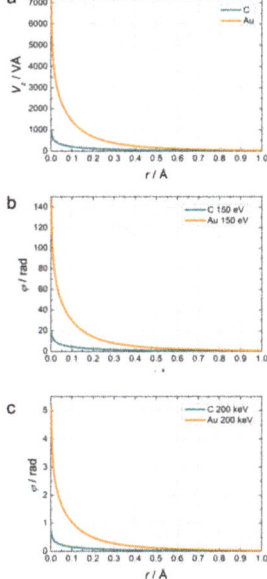

Figure 5. Calculated projected potentials for C and Au atoms (**a**), relative phase shifts introduced by C and Au atoms when probed with 150 eV (**b**), and 200 keV (**c**) electrons.

2.4.3. Phase Shift of an Electron Wave in a Magnetic Potential

For a charged particle moving in a magnetic potential, the total momentum is given by:

$$\vec{p} = \frac{m\vec{v}}{\sqrt{1-v^2/c^2}} + q\vec{A}. \tag{29}$$

This momentum is preserved during the movement of the particle in a magnetic potential. The phase shift of an electron moving in a magnetic potential can be written in the form

$$\varphi_M(x,y) = -\frac{e}{\hbar} \int_{-\infty}^{+\infty} A_z(x,y) dz. \tag{30}$$

The phase difference between two arbitrary points at coordinates (x_1, y_1) and (x_2, y_2) can be written in the form of a loop integral

$$\Delta\varphi_M = \varphi_M(x_1, y_1) - \varphi_M(x_2, y_2) = -\frac{e}{\hbar} \oint \vec{A} \, d\vec{l} \tag{31}$$

for a rectangular loop formed by two parallel electron trajectories crossing the sample at coordinates (x_1, y_1) and (x_2, y_2) and joined, at infinity, by segments perpendicular to the trajectories [10]. By applying Stoke's theorem, the phase shift can be expressed through the magnetic flux:

$$\Delta\varphi_M = \frac{e}{\hbar} \int \vec{B} \, d\vec{S} = \frac{e}{\hbar} \Phi_M, \tag{32}$$

where Φ_M is the magnetic flux through the whole region of space bounded by two electron trajectories crossing the sample at the positions of these two points.

2.5. Wavefront Propagation: Fresnel and Fraunhofer Diffraction

The propagation of electron waves can be described by the diffraction theory. We suppose that the complex-valued wavefront distribution is known at some plane (ξ, η). The propagation of a complex-valued wavefront to a point P_0 in the plane (x, y) can be calculated by employing the Huygens-Fresnel principle [11]:

$$u_0(P_0) = -\frac{i}{\lambda} \iint_S u_1(P_1) \frac{e^{ikr_{01}}}{r_{01}} dS, \tag{33}$$

where P_1 is a point in the plane (ξ, η), r_{01} is the distance between points P_0 and P_1, and the integration is performed over the entire plane (ξ, η), as illustrated in Figure 6.

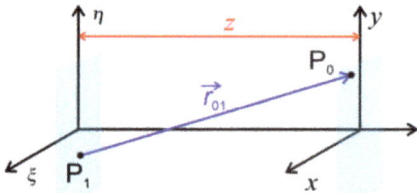

Figure 6. Schematic of symbols used in the Huygens-Fresnel principle, and the Fresnel and Fraunhofer diffraction integrals.

r_{01} can be written using the Taylor series, as follows:

$$r_{01} = \sqrt{(x-\xi)^2 + (y-\eta)^2 + z^2} \approx z + \frac{(x-\xi)^2 + (y-\eta)^2}{2z}, \quad (34)$$

provided that

$$z^3 \gg \frac{\pi}{4\lambda}\left[(x-\xi)^2 + (y-\eta)^2\right]^2_{\max}. \quad (35)$$

At a distance z which satisfies Equation (35), the Fresnel diffraction regime is observed, and the diffracted wavefront is described by

$$U_0(x,y) \approx -\frac{i}{\lambda z}e^{ikz}\iint_S U_1(\xi,\eta) e^{ik\frac{(x-\xi)^2+(y-\eta)^2}{2z}} d\xi d\eta, \quad (36)$$

which is obtained by substituting Equation (34) into Equation (33). From Equation (36), we see that in the Fresnel diffraction regime, the distribution of the propagated wave is simply given by a convolution of the original wave distribution with the free space propagation function $e^{ik\frac{x^2+y^2}{2z}}$.

At even larger z distances, such that

$$z \gg \frac{\pi}{\lambda}\left[\xi^2 + \eta^2\right]_{\max}, \quad (37)$$

a Taylor series expansion gives

$$r_{01} \approx z + \frac{(x^2 - 2x\xi) + (y^2 - 2y\eta)}{2z}, \quad (38)$$

and by substituting Equation (38) into Equation (33), we obtain the wavefront distribution in the Fraunhofer diffraction regime

$$U_0(x,y) \approx -\frac{i}{\lambda z}e^{ikz}e^{ik\frac{x^2+y^2}{2z}}\iint_S U_1(\xi,\eta) e^{-\frac{2\pi i}{\lambda z}(x\xi+y\eta)} d\xi d\eta, \quad (39)$$

which is just a two-dimensional (2D) FT of the wavefront distribution at the plane (ξ, η).

3. Holography Principle

In general, holography can be described as a measurement technique in which a known signal is superimposed with an unknown signal, and the latter can then be unambiguously reconstructed from the interference pattern that is created. For two interfering waves, this principle can be mathematically written as the holographic equation:

$$H = |U_R + U_O|^2 = |U_R|^2 + |U_O|^2 + U_R^* U_O + U_R U_O^*, \quad (40)$$

where U_R is the reference wave, U_O is the object (unknown) wave, and H is the resulting hologram. In Equation (40), the first term $|U_R|^2$ is a constant distribution associated with the background, which is obtained without the presence of the object. The second term $|U_O|^2$ is assumed to be small, and can be neglected. The third and fourth terms are the object and twin image terms, $U_R^* U_O$ and $U_R U_O^*$, respectively, which create the interference pattern. From the holographic equation, it follows that provided H and U_R are known, U_O can be reconstructed as $U_R H \propto U_O + U_R^2 U_O^*$. However, there will be always a remaining signal from the conjugated twin image term, $U_R^2 U_O^*$.

4. Coherence

All imaging techniques that employ an interference pattern (for example holography or coherent diffractive imaging (CDI)) require coherent waves. Coherence characterizes the stability of the phase difference between two interfering waves. The contrast of an interference pattern is given by the coherence of the interfering waves. The visibility (contrast) of the interference pattern gives the degree of coherence [12,13]. Probing radiation is characterized by temporal (longitudinal) and spatial (transverse) coherence.

Temporal (longitudinal) coherence is a measure of how monochromatic a source is. The temporal coherence length $l_c^{Temporal}$ of a source with wavelength spread $\lambda \pm \Delta\lambda$ is given by $l_c^{Temporal} \approx \lambda^2/\Delta\lambda$. For example, $l_c^{Temporal} \approx 390$ nm for low-energy electrons with energy 250 ± 0.1 eV, and $l_c^{Temporal} \approx 1$ μm for high-energy electrons with energy $200,000 \pm 1$ eV. In both cases, the temporal coherence length exceeds the sizes of the objects typically studied in electron microscopy.

Spatial (transverse) coherence is defined by the size of the virtual source. According to the van Cittert-Zernike theorem [14,15], the complex coherence factor is given by the FT of the intensity distribution of the source [16]. For a source with intensity distribution described by a Gaussian function $s(\xi,\eta) = \exp\left(-\frac{\xi^2+\eta^2}{2\sigma^2}\right)$, where (ξ,η) are the coordinates in the source plane and σ is the standard deviation, the spatial coherence length at a distance L from the source is given by $l_c^{Spatial} = \frac{\lambda L}{2\pi\sigma}$ [17]. The spatial coherence is inversely proportional to the source size. For example, for low-energy electrons of energy 250 eV (wavelength = 0.078 nm), and a virtual source with $\sigma = 1$ Å, the coherence length amounts to about 120 nm at a distance of $L = 1$ μm from the source, which is sufficient to image a macromolecule of a size of few tens of nanometers. For high-energy electrons of energy 200 keV (wavelength = 2.51 pm), and a virtual source with $\sigma = 1$ Å, the coherence length amounts to about 4 nm at a distance of $L = 1$ μm from the source. It must be noted that in the case of high-energy electrons employed in a TEM, the spatial coherence length depends not only on the source properties but it also scales with the beam size as the beam propagates in TEM [18], being typically a few tens of nanometers.

5. Principle of Gabor Holography

The first electron microscope was built by Ernst Ruska and Max Knoll between 1931 and 1933 [19]. The very short wavelengths of electrons gave rise to the hope that these microscopes could be used to visualize very small objects such as viruses. Although biologists had already identified the function and activity of viruses before the era of electron microscopy [20,21], their geometrical shapes remained a mystery, and with the invention of the electron microscope, it become possible to visualize these shapes. The first images of viruses were obtained by Gustav Kausche, Edgar Pfankuch, and Helmut Ruska in 1939 using an electron microscope; they imaged a tobacco mosaic virus (TMV) and identified its simple geometrical shape, a rod 18 nm in diameter and 300 nm in length [22], as shown in Figure 7. TMV has a remarkable history of "firsts", since it was also the first virus to be discovered and named [21].

Figure 7. Images of the tobacco mosaic virus obtained by Gustav Kausche, Edgar Pfankuch, and Helmut Ruska in 1939, using an electron microscope. Reprinted from [22] by permission from Springer Nature, copyright 1939.

However, these electron microscope images of TMVs (also shown in Figure 7) had low resolution and quality. The question therefore arose as to why the resolution of the image was so low, despite the very short wavelength of the electrons used, which was much smaller than the typical interatomic distance of 1.5–2 Å, and this issue occupied many scientists at the time. In 1936, Otto Scherzer published his research on the subject (now known as the Scherzer theorem), in which he studied the properties of an imaging system using electromagnetic lenses. Scherzer demonstrated that because of the symmetry of the electromagnetic lenses, static electromagnetic fields and the absence of space charges (properties of a typical electromagnetic lens system), aberrations (mainly chromatic and spherical) will always be present and will degrade the resolution, thus preventing atomic resolution images [23]. Scientists therefore started to search for solutions to the aberrations problem. Eventually, aberration-corrected transmission electron microscopes (ACTEM) were developed in 2010, which delivered images with atomic resolution. However, there was also another outstanding solution to the aberrations problem, which resulted in a completely new technique—holography.

In 1947, Dennis Gabor patented a novel imaging technique which he named "holography" [24–26]. Gabor's ingenious idea for solving the aberrations problem in an electron microscope was to remove all the lenses between the sample and the detector. He argued that since the electron wave that was partially scattered by the sample (object wave) would interfere with the unscattered (reference) wave, the resulting interference pattern formed in the detector plane would contain the complete information about the object wave, and the entire object distribution could therefore be reconstructed. Although Gabor envisioned this new type of microscopy being applied in an electron microscope (Figure 8a,c), he proved this principle using an optical experiment (Figure 8b,d) [25,26]. This form of holography is called Gabor-type or in-line holography, since the object and the reference wave share the same optical axis.

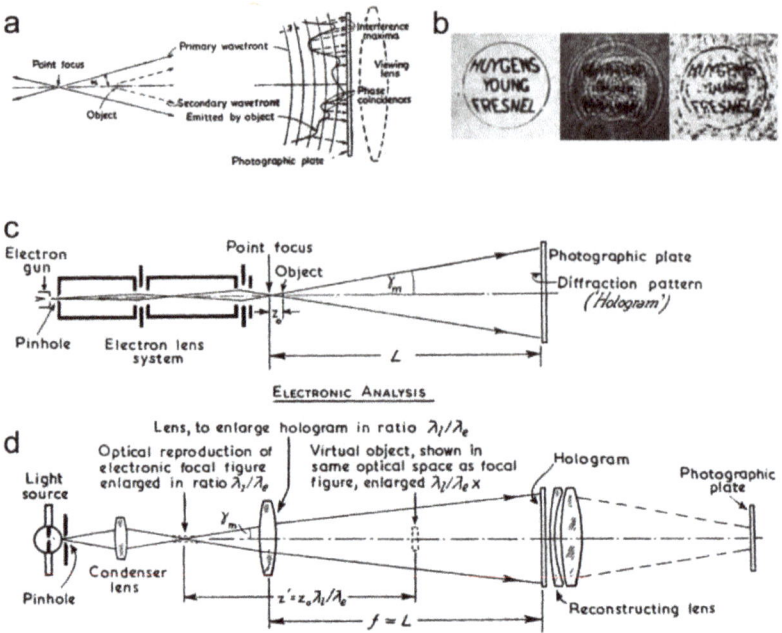

Figure 8. Principle of holography, as illustrated by Dennis Gabor [25,26]. (**a**,**c**) schematic of realization of holography in a transmission electron microscope. (**b**) Sample (left) with three words written on a transparent film, its hologram (middle) recorded on photographic film, and the reconstructed hologram (right) as a result of optical holography experiments involving recording and reconstruction of holograms, as shown in (**d**). (**a**,**b**) Reprinted from [25] by permission from Springer Nature, copyright 1948.

6. Point Projection Microscopy (PPM)

A frequently used experimental scheme that is similar to in-line holography is called point projection microscopy (PPM). The reason for these different names for two almost identical experimental arrangements is as follows. In 1939, George A. Morton and Edward G. Ramberg published a half-page article entitled "Point projector electron microscope," in which they described a novel type of electron microscopy [27]. Their technique employed "an etched tungsten or molybdenum point" cathode as an electron source, a specimen that was partially transparent, and no lenses between the sample and the detector. The image formed on the detector was a magnified image of the sample, where the magnification was determined by the ratio between the source-to-detector and the source-to-sample distances. They produced experimental images of a copper grid, as shown in Figure 9. In their experiments, Morton and Ramberg achieved a magnification of up to 8000 times, but did not publish their images at this magnification because, as they explained, the quality of the images was degraded because of insufficient mechanical steadiness. However, it is possible that this reduction in the quality of the images (that is, the degraded sharpness of the edge) was in fact caused by the diffraction and interference effects which arise at shorter source-to-sample distances. Thus, Morton and Ramberg may have observed the first holograms already in 1939.

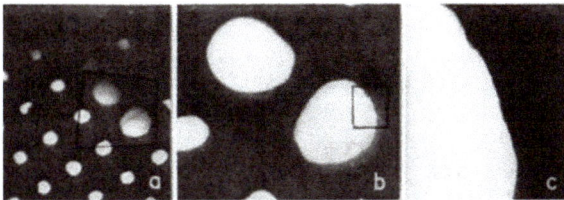

Figure 9. Images of a copper grid obtained by Morton and Ramberg using point projection microscopy (PPM) with electrons, at magnifications of (**a**) 200, (**b**) 600, and (**c**) 3000 times. Figure reprinted from [27], copyright (1939) by the American Physical Society.

The main difference between PPM and in-line holography is that in the former, the resulting image is a projection image rather than an interference pattern (Figure 10a), while in the latter, the resulting image is an interference pattern formed by interference between the scattered and non-scattered waves (Figure 10b). Thus, the same experimental setup can be utilized in two regimes, since at shorter source-to-sample distances, an interference pattern emerges and a point projection image turns into a hologram. This is why this experimental arrangement is referred to by some researchers as PPM [28–36], and by others as in-line holography [37,38]. Another point which should be mentioned is that in the in-line holography proposed by Gabor, the point source is a focused spot, rather than a physical source as in PPM.

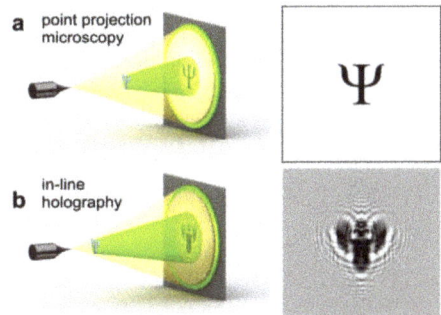

Figure 10. Point projection microscopy (PPM) and in-line holography. (**a**) Experimental arrangement for PPM and the resulting image. (**b**) Experimental arrangement for in-line holography and the resulting image.

7. Off-Axis Holography

7.1. The Electron Biprism

In 1956, Möllenstedt and Düker invented a method for splitting electron beams by placing a positively charged wire within the electron wave, orthogonal to the propagation of the wavefront. In this scheme, electrons passing the positively charged wire are deflected toward the wire, and thus the electron wave is split into two overlapping wavefronts, creating an interference pattern of equidistant fringes, as illustrated in Figure 11. The electron biprism thus acts in an analogous way to an optical prism (hence the term "biprism") [39].

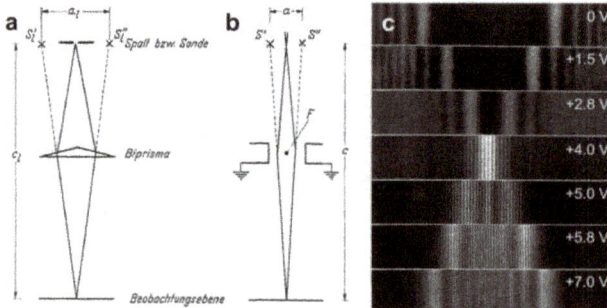

Figure 11. A biprism in an electron microscope. Ray diagrams for (**a**) an optical prism and (**b**) an electron biprism in an electron microscope. (**c**) Experimentally recorded electron biprism interference patterns with different potentials applied to the biprism, using a wire 2 µm in diameter. Reprinted from [39] by permission from Springer Nature, copyright 1956.

7.2. Measuring Potentials Using Off-Axis Holography

7.2.1. Electrostatic Potential

In 1957, one year later after demonstrating the principle of the electron biprism, Möllenstedt and Keller placed a sample into the one of the two split electron beams and measured the resulting interference pattern, thus creating the first off-axis electron hologram [40]. Their experimental arrangement is shown in Figure 12a. Based on the acquired interference pattern, they were able to measure the electrostatic potential of a sample as follows. The sample consisted of strips of carbon film of different thicknesses, 40 and 160 Å, and the thickness was measured by optical absorption. The accelerating voltage was $U = 54.4$ kV. The phase shift caused by different potentials was evaluated as $\Delta\varphi = 2\pi\left(\frac{D}{\lambda_m} - \frac{D}{\lambda_v}\right) = -2\pi D \frac{\Delta\lambda}{\lambda^2}$ where λ_v is the wavelength in a vacuum, λ_m is the wavelength in the material, D is the difference in thickness, and here $D = 120$ Å. $\lambda = \sqrt{\frac{150}{U}}$ (λ in Å, U in volts), thus giving $\Delta\lambda = -\frac{1}{2}\sqrt{\frac{150}{U^3}}\Delta U$. From the bending of the fringes in the interference pattern, which correspond to the regions of different sample thickness (as shown Figure 12b), Möllenstedt and Keller evaluated the phase shift to be about $\Delta\varphi = \pi \pm 15\%$, and calculated the potential as $\Delta U = \frac{300\Delta\varphi}{2\pi D\lambda} = (24 \pm 5)$ V [40]. Since this first experiment, the measurement of electrostatic potentials has been one of the main applications of electron off-axis holography [41–43]. Recently, high-energy off-axis electron holography has been applied for imaging individual charges and the electrostatic charge density distributions with a precision of better than a single elementary charge [44,45].

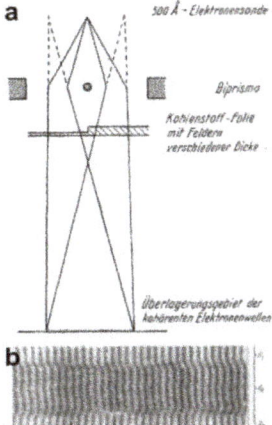

Figure 12. Off-axis electron holography. (**a**) Experimental arrangement. (**b**) Off-axis electron hologram exhibiting the shift in the interference pattern caused by the different thicknesses of the sample, and therefore by the differences in potential and the additional phase shift. Reprinted from [40] by permission from Springer Nature, copyright 1957.

7.2.2. Magnetic Potential

The possibility of measuring the magnetic potential was proposed shortly after the above demonstration of measuring the electric potential using off-axis electron holography. In 1959, Yakir Aharonov and David Bohm published a theoretical paper in which they described a quantum mechanical phenomenon whereby an electrically charged particle will be affected by an electromagnetic potential, despite being confined to a region in which both the magnetic field B and electric field E are zero [46]. This phenomenon is known now as the Aharonov-Bohm (AB) effect, and since it cannot be explained in the frame of classical electrodynamics, it is a truly quantum phenomenon. In their paper, Aharonov and Bohm even provided a sketch of an experimental scheme based on electron interference, which was an arrangement very similar to off-axis holography. In 1960, the corresponding experiment was conducted by Chambers [47], who demonstrated a shift in an electron interference pattern caused by an enclosed magnetic flux, thus proving the AB effect and the quantum nature of the electronic interaction with the magnetic potential. Nowadays, electron off-axis holography is routinely applied in measuring the magnetic properties of material science and biological samples in transmission electron microscopes operating with keV energy electrons [10,48–52].

7.3. Reconstruction of an Off-Axis Hologram

Superimposing a tilted reference wave U_R and an object wave U_O, where

$$U_R(\vec{r}) = \exp(i\vec{q}_R \vec{r}),$$
$$U_O(\vec{r}) = A_O(\vec{r}) \exp[i\varphi_O(\vec{r})],$$
(41)

yields an interference pattern with

$$I(\vec{r}) = 1 + |A_O(\vec{r})|^2 + 2A_O(\vec{r}) \cos[\vec{q}_R \vec{r} + \varphi_O(\vec{r})],$$
(42)

where the tilt of the reference wave is specified by the two-dimensional reciprocal space vector $\vec{q}_R$, $A_O(\vec{r})$ and $\varphi_O(\vec{r})$ refer to amplitude and phase, respectively. From Equation (42), it can be shown that the FT of a hologram can be written in the form

$$FT(I) = \delta(\vec{q}) + FT(|A_O|^2) + \delta(\vec{q} + \vec{q}_R)FT[A_O\exp(-i\varphi_O)] + \delta(\vec{q} - \vec{q}_R)FT[A_O\exp(i\varphi_O)]. \qquad (43)$$

The resulting 2D complex-valued Fourier spectrum consists of the autocorrelation (central band) and two mutually conjugated sidebands centered at the carrier frequencies ($\vec{q}_R$ and $-\vec{q}_R$), as shown in Figure 13b.

The numerical reconstruction of an off-axis hologram (recorded with light, electrons, or any other radiation) consists of the following steps, and an example is shown in Figure 13. (i) A 2D FT of the hologram (Figure 13a) is calculated. (ii) In the resulting 2D spectrum (Figure 13b), one of the two sidebands is selected by applying a low-pass filter centered on the chosen sideband, setting the central band and the other sideband to zero. (iii) The selected sideband is shifted to the center of the spectrum. (iv) The resulting complex-valued spectrum is then inverse Fourier transformed back to the real space. (v) The 2D amplitude (given by $A_O(\vec{r})$) and phase (given by $\varphi_O(\vec{r})$) distributions are extracted from the obtained distribution, as shown in Figure 13c,d.

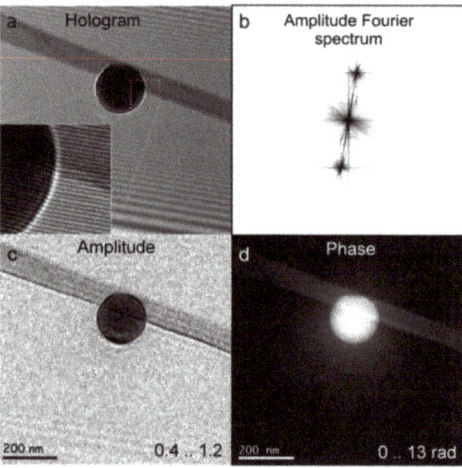

Figure 13. Electron off-axis hologram of a latex sphere and its reconstruction. (**a**) Off-axis hologram of a latex sphere recorded at 200 keV, with Fresnel fringes from the biprism filament edge readily visible. (**b**) Amplitude of the Fourier spectrum of the hologram shown in (**a**). (**c**) Reconstructed amplitude. (**d**) Unwrapped reconstructed phase, with phase values between 0 and 13 rad. Reprinted from [53], with permission from Elsevier.

7.4. Low-Energy Electron Off-Axis Holography

Off-axis holography in a low-energy electron microscope has been demonstrated by Roger Morin and colleagues, and has been reported in several publications [54–58]. A schematic of the experimental arrangement is similar to the PPM or Gabor in-line holography, shown in Figure 14a. The sample is placed into a divergent electron wave, and the biprism is placed between the sample and the detector. An electrostatic lens is used to magnify the image of the obtained interference pattern. Since PPM and Gabor in-line holography do not use any lenses, the image of the sample is always a defocused (due to Fresnel diffraction) image of the sample. Morin and colleagues reported a series of experiments imaging carbon foil (an example is shown in Figure 14b–d) [54–58]; they also imaged a sharp magnetic

(Ni) tip above and below the Curie temperature and observed the phase shift related to the magnetic flux [58].

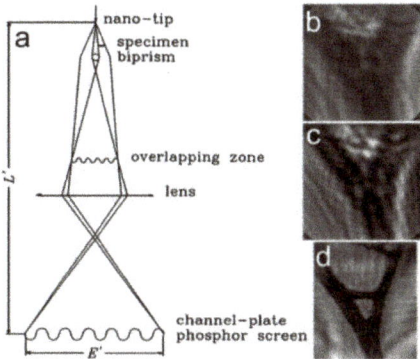

Figure 14. Low-energy electron off-axis holography. (**a**) Schematic arrangement of the low-energy holographic electron microscope with a biprism. (**b–d**) Imaging of a perforated carbon foil: (**b**) off-axis hologram, (**c**) in-line hologram without biprism, and (**d**) amplitude reconstructed from (**b**). The field of view is 217 nm. Figure reprinted from [56], copyright (1996) by the American Physical Society.

7.5. Further Reading about Off-Axis Holography

For further reading about off-axis electron holography, numerous papers by Hannes Lichte and his colleagues can be recommended as tutorials [13,59–61]. The applications of off-axis electron holography in materials science are discussed in references [51,62], and for biological samples in references [63,64]. The performance limits of off-axis holography are discussed by Lichte in reference [65].

8. In-Line Holography

8.1. In-Line Holography in TEM

8.1.1. Defocused, Over-Focused, and Under-Focused Imaging

In-line holography is easily realized in TEM by simply defocusing the image of the sample. In this case, the interference between the unscattered and scattered waves forms the in-line hologram. A particularly interesting case is that of phase objects. Most biological macromolecules such as proteins are composed of atoms (C, H, O, N) which are relatively weak scatterers, so that the entire macromolecule is a weak phase object and creates no contrast when imaged in focus. The phase object only causes significant contrast when imaged in defocus.

A defocused image can be obtained in the over- or under-focused regime, as illustrated in Figure 15a–c. This change in contrast can be explained using the transport of intensity (TIE) equation [66]:

$$\frac{\partial I(x,y,z)}{\partial z} + \nabla_{x,y}\left[I(x,y,z)\frac{\nabla_{x,y}\varphi(x,y)}{k}\right] = 0 \tag{44}$$

where $I(x,y,z)$ is the intensity and $\varphi(x,y)$ is the phase distribution at a plane z. For a phase object imaged in focus, the wavefront is given by $U(x,y,z) = \exp[i\varphi(x,y)]$, the intensity $I(x,y,z) = 1$ and Equation (44) becomes

$$\frac{\partial I(x,y,z)}{\partial z} + \frac{1}{k}\nabla_{x,y}^2 \varphi(x,y) = 0. \tag{45}$$

By replacing the differential with a numerical differentiation, as $\frac{\partial I(x,y,z)}{\partial z} \approx \frac{I(x,y,z+\Delta z)-I(x,y,z)}{\Delta z}$, we can rewrite the TIE Equation (45) as

$$I(x,y,z+\Delta z) = 1 - \frac{\lambda \Delta z}{2\pi} \nabla^2_{x,y} \varphi(x,y). \tag{46}$$

In Equation (46), the intensity distribution is proportional to the second derivative of the phase distribution.

A simulated example is shown in Figure 15d–h. Here, the test object is a hole in a thin carbon film that causes a phase shift of 1 radian. From experimental observations, it is known that the under- and over-focused images of such a sample have clear signatures: the edge of the hole exhibits a bright fringe in the under-focused image, and a dark fringe in the over-focused one. The second derivative of the phase distribution is shown in Figure 15h. For the under-focused image, $\Delta f > 0$ and $\Delta z < 0$, and according to Equation (46), $I(x,y,z+\Delta z) \propto \nabla^2_{x,y} \varphi(x,y)$, giving a bright fringe at the edge of the hole. For the over-focused image, $\Delta f < 0$ and $\Delta z > 0$, and according to Equation (46), $I(x,y,z-\Delta z) \propto -\nabla^2_{x,y} \varphi(x,y)$, meaning that a dark fringe is seen at the edge of the hole.

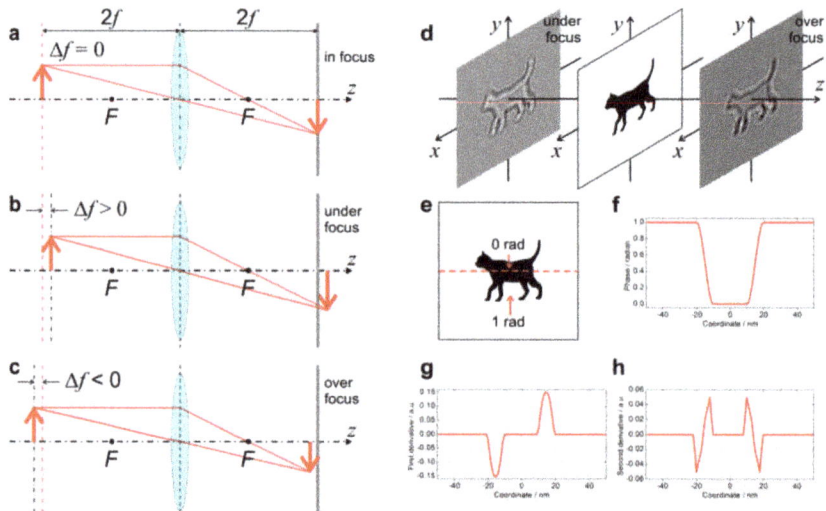

Figure 15. Defocused imaging in TEM. (**a**–**c**) Ray diagrams of a lens system when imaging (**a**) in focus, (**b**) under focus, and (**c**) over focus. The detector is at the same position in all three cases. The position of the object is shifted along the optical axis in (**b**) by $\Delta f > 0$ and in (**c**) by $\Delta f < 0$, and as a result, the image on the detector appears under-focused in (**b**) and over-focused in (**c**). (**d**) Under- and over-focused images of a pure phase object, a cat-shaped hole in a carbon film with the phase distribution shown in (**e**). (**f**–**h**) show profiles through the middle of the 2D distribution of (**f**) the sample, (**g**) its first derivative, and (**h**) its second derivative.

8.1.2. Focal (Defocus) Series

In 1986, Schiske proposed the possibility of full wavefront reconstruction from a sequence of intensity measurements acquired at different defocus distances in an electron microscope [67]. In 1992, Coene et al. demonstrated the unambiguous high-resolution reconstruction of samples obtained from a focal series acquired in a TEM, and this has become a practical tool for image analysis in high-resolution transmission electron microscopy (HRTEM) [68]. HRTEM images of material science samples (and particularly crystals) often display misleading contrast; for example, bright spots can be mistaken for atoms but in reality are the spaces between atoms. Strictly speaking, HRTEM images cannot be interpreted alone, and corresponding simulations must be performed to match the experimental images. For an unambiguous determination of the structure, focal image series can be acquired and reconstructed using numerical procedures, thus recovering the complex-valued exit wave at atomic resolution [69]. Focal series of images can be reconstructed by employing the TIE [66], as has been demonstrated for light optical [70] and electron holograms [71] or by iterative phase retrieval methods. A sequence of in-line electron holograms acquired at different defocus distances and their reconstruction, obtained using the flux-preserving iterative reconstruction algorithm described in [71], are shown in Figure 16.

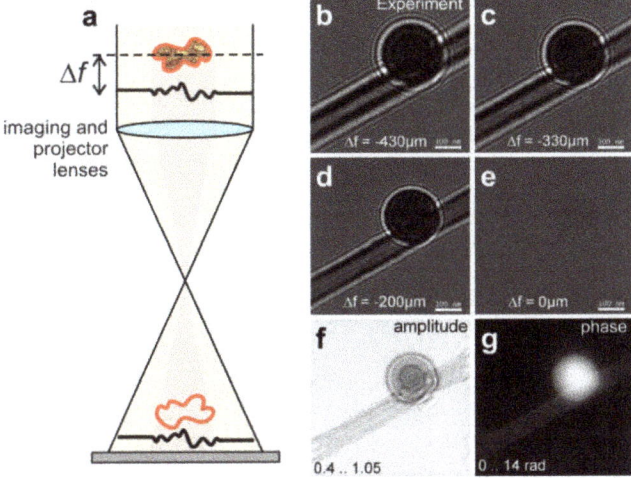

Figure 16. Defocus series in transmission electron microscope (TEM). (**a**) Drawing of a holographic in-line scheme, where the red (yellow) color represents the object (reference) wave and Δf is the defocus distance. (**b–e**) Experimental in-line holograms of a latex sphere recorded at different values of defocus. (**f**) Amplitude and (**g**) phase of the object wave, reconstructed using an iterative flux-preserving focal series reconstruction algorithm [71]. Reprinted from [53], with permission from Elsevier.

8.1.3. Single In-Line Hologram and Its Reconstruction

The object distribution can be also reconstructed from a single in-line hologram (defocus image) by applying iterative reconstruction, as explained in detail in the literature [53,72–76]. We provide only a brief summary of the reconstruction steps here. Before reconstruction, the hologram is divided by the background image, which is recorded under exactly the same experimental conditions as the hologram, only without the object. Alternatively, the background image can be created numerically by simple low-pass filtering of the hologram of the object. The hologram divided by the background image is the normalized hologram. The distribution of this normalized hologram does not depend on the parameters such as the intensity of the reference wave, and is described mathematically by the holographic equation with a reference wave of amplitude one. The normalized hologram can be reconstructed as described elsewhere [77], and quantitatively correct absorption and phase distributions of the sample can be extracted [78]. Next, an iterative reconstruction routine is applied based on the Gerchberg-Saxton algorithm [79], in which the wavefront is propagated back and forth between the two planes (i.e. the hologram and object planes), and constraints are applied in each plane. In the hologram plane, the updated amplitude is replaced with the square root of the measured intensity, while in the object plane, support constraint [80,81], positive absorption constraint [72], and/or real and positive constraints can be applied to the reconstructed object distribution. An example of a latex sphere reconstructed from a single in-line electron hologram (obtained by defocused imaging in a

TEM) with positive absorption and finite support constraints is shown in Figure 17. Here, the sample exhibits absorption and a significant phase shift [53].

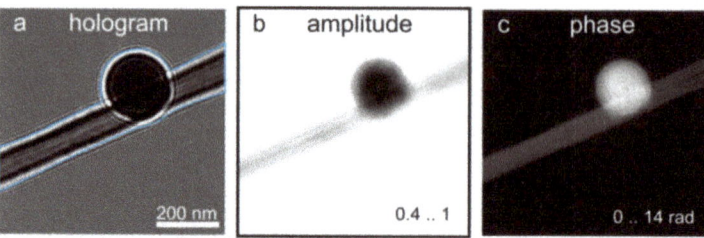

Figure 17. In-line hologram of a latex sphere and its reconstruction. (**a**) In-line electron hologram of the latex sphere recorded at the defocus 180 μm, with 200 keV electrons in a TEM. The blue lines mark the area outside of which the transmission was set to 1 during the iterative reconstruction (support). (**b**,**c**) show the retrieved amplitude and phase distributions, respectively. Reprinted from [53], with permission from Elsevier.

8.2. Low-Energy Electron Holography

8.2.1. Experimental Arrangement

An experimental arrangement for in-line holography with low-energy electrons [37,82] is sketched in Figure 18. Electrons are extracted by field emission from a sharp tungsten W (111) tip, with energy 30–250 eV. A sample is placed in front of the tip at a distance d from the source, where d ranges from tens of nanometers to a few microns. The in-line hologram formed by the interference between the scattered and non-scattered wave is acquired by a detector positioned at a distance D from the source, D is typically 5–20 cm. The magnification of the microscope is given by the ratio D/d. The technical details of low-energy holographic microscopes are provided in references [37,83,84].

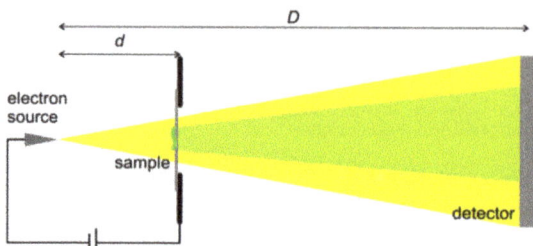

Figure 18. Experimental arrangement for in-line holography with low-energy electrons.

8.2.2. Reconstruction of In-Line Holograms

Algorithms for the simulation and reconstruction of in-line holograms are provided and explained in detail in reference [77]. Here, we discuss the main conclusions of the theory of formation and reconstruction of in-line holograms.

Plane waves. In-line holography is often realized with plane waves. In this case, the incident wavefront is a plane wave, and the interference pattern (hologram) is formed at some not too far distance from the sample. The distribution of the interference pattern changes with the distance from the sample. The complex-valued wavefront at the detector is given by the Fresnel diffraction integral, and in the paraxial approximation can be calculated as a convolution:

$$U_{\text{detector}}(X,Y) = -\frac{i}{\lambda} \iint_S t(x,y) \exp\left\{\frac{i\pi}{\lambda z}\left[(x-X)^2 + (y-Y)^2\right]\right\} dx dy$$
$$\propto t(X,Y) \otimes \exp\left[\frac{i\pi}{\lambda z}(X^2 + Y^2)\right] \quad (47)$$

where $t(x,y)$ is the transmission function of the sample, (x,y) are the coordinates in the sample plane, (X,Y) are the coordinates in the detector plane, z is the distance between the sample and the detector, and $\otimes$ denotes convolution.

Spherical waves. The original in-line Gabor-type holography employed a divergent incident wavefront. In this arrangement, even though the resulting interference pattern (hologram) is acquired in the far field, the distribution of the diffracted wave is described by Fresnel diffraction. The interference pattern has the same appearance at any distant detecting plane, and moving the detecting plane away from the sample will result only in an increased magnification of the interference pattern. However, changing the distance between the source and the sample will change the distribution of the interference pattern (hologram); that is, it will have the same effect as changing the sample-to-detector distance in in-line holography with plane waves.

The complex-valued wavefront at the detector plane is given by the Fresnel diffraction integral, and in the paraxial approximation can be calculated as a convolution:

$$U_{\text{detector}}(X,Y) = -\frac{i}{\lambda} \iint_S \exp\left[\frac{i\pi}{\lambda z_1}(x^2+y^2)\right] t(x,y) \, \exp\left\{\frac{i\pi}{\lambda(z_2-z_1)}\left[(x-X)^2+(y-Y)^2\right]\right\} dx dy$$
$$\propto \iint_S t(x,y) \exp\left\{\frac{i\pi}{\lambda z_1}\left[\left(x-\frac{z_1}{z_2}X\right)^2 + \left(y-\frac{z_1}{z_2}Y\right)^2\right]\right\} dx dy, \quad (48)$$

where z_1 is the distance between the source and the sample, and z_2 is the distance between the source and the detector. By introducing the scaled coordinates $X' = \frac{z_1}{z_2}X = \frac{X}{M}$ and $Y' = \frac{z_1}{z_2}Y = \frac{Y}{M}$, where $M = \frac{z_2}{z_1}$ is the magnification factor, Equation (48) can be written as:

$$U_{\text{detector}}(X',Y') \propto o(X',Y') \otimes \exp\left[\frac{i\pi}{\lambda z_1}(X'^2 + Y'^2)\right] \quad (49)$$

which implies that a hologram recorded with a spherical wave with source-to-sample distance z_1 can be treated as a hologram recorded with a plane wave with sample-to-detector distance z_1, and the coordinates are scaled by a magnification factor M.

For a thin sample that can be approximated by a 2D distribution in one plane, a hologram acquired with a spherical wave can be reconstructed as if it had been obtained with a plane wave, as illustrated in Figure 19. The following relation holds [77]:

$$\frac{\lambda z}{S_{\text{plane}}^2} = \frac{\lambda z_2^2}{z_1 S_{\text{spherical}}^2} \quad (50)$$

where $S_{\text{plane}} \times S_{\text{plane}}$ and $S_{\text{spherical}} \times S_{\text{spherical}}$ are the sizes of the hologram recorded with plane and spherical waves, respectively.

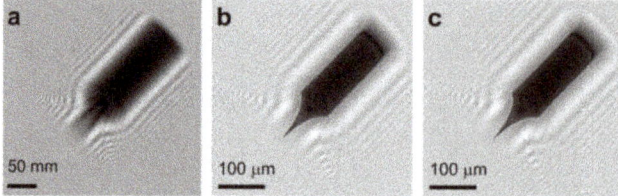

Figure 19. Optical hologram of a tungsten tip and its reconstruction. (**a**) Hologram recorded with 532 nm laser light in an in-line Gabor scheme with spherical waves, with a source-to-sample distance of 1.4 mm, and a source-to-screen distance of 1060 mm. (**b**) Amplitude of the object distribution reconstructed from the hologram shown in (**a**) using the reconstruction algorithm for spherical waves, where the size of the reconstructed area is 429 × 429 µm^2. (**c**) Amplitude of the object distribution from the hologram shown in (**a**) using the reconstruction algorithm for plane waves, assuming a hologram size of 429 × 429 µm^2 and a sample-to-hologram distance of 1.4 mm. Adapted from [77].

8.2.3. Imaging Biological Samples and Individual Macromolecules

Low-energy electrons with kinetic energies in the range 30–250 eV have the advantage of causing no significant radiation damage to biological molecules; this has been exemplified by the continuous exposure of individual DNA molecules to low-energy electrons for 70 min, without a noticeable change in their in-line holograms at a resolution of 1 nm [85,86]. The number of electrons required to acquire a single 20 ms low-energy electron hologram at a resolution of 1 nm amounts to about 250 electrons per 1 Å^2, which translates into a radiation dose of 4.58×10^{11} Gray.

Low-energy electron in-line holography has been successfully applied to the imaging of various individual biological molecules, for example purple protein membrane [84], DNA molecules [30,38,86,87], phthalocyaninato polysiloxane molecule [28], TMV [88,89], a bacteriophage [90], ferritin [91], and individual proteins (bovine serum albumin, cytochrome C, and hemoglobin) [92]; some of these results are shown in Figures 20 and 21.

8.2.4. Imaging Electric Potentials

Local electric potentials such as those created by individual charged adsorbates on graphene [93] can be visualized using low-energy electron in-line holography [94], with a sensitivity of a fraction of an elementary charge. Some results are shown in Figure 22. Low-energy electrons exhibit a sensitivity that is hundreds of times higher to local electric potentials than high-energy electrons. This can be explained intuitively based on the fact that an electron moving at a lower speed spends more time in the potential, and thus gets deflected more. An adsorbate with one elementary charge can cause about 30% and 1% contrast in an in-line hologram acquired with low- and high-energy (100 keV) electrons, respectively (Figure 22e,f).

Iterative reconstruction of in-line holograms of individual charges provides the amplitude (associated with absorption) and phase (associated with the potential) distributions [74] (Figure 23). The reconstructed absorption distributions (Figure 23d,e) appear to be narrower than the reconstructed phase distributions (Figure 23f,g). This agrees well with the notion that the phase distribution (unlike the absorption distribution) does not reflect the actual size of the adsorbate, but instead reflects the potential distribution caused by the charge.

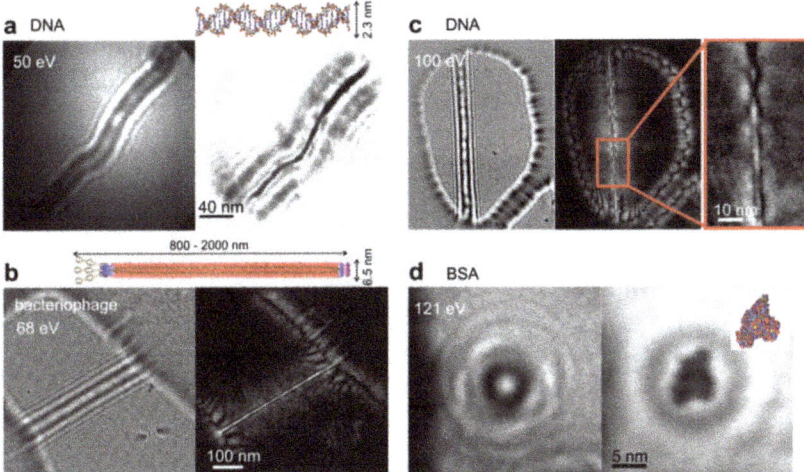

Figure 20. Low-energy in-line holography imaging of individual macromolecules, showing results obtained by Fink et al. University of Zurich. In each case, the left image shows experimental holograms and the right shows the corresponding reconstructions: (**a**) DNA molecules, copyright OSA 1997 [38], (**b**) bacteriophage molecule (reprinted by permission from Springer Nature [90], copyright 2011), (**c**) DNA molecule [87] (copyright Springer Nature 2013), and (**d**) bovine serum albumin (BSA) molecules [92].

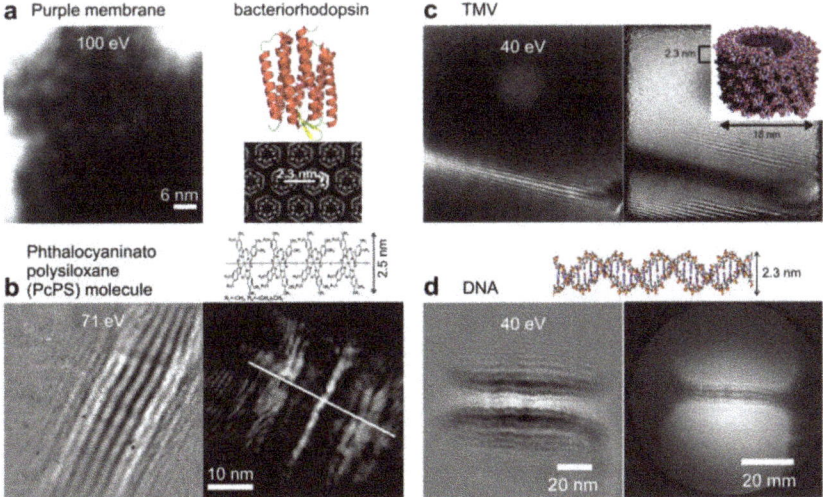

Figure 21. Low-energy in-line holography imaging of individual macromolecules. In each case, the left image shows experimental holograms, and the right shows the corresponding reconstructions. (**a**) Purple membrane (reprinted from [84], with permission from Elsevier). (**b**) Phthalocyaninato polysiloxane (PcPS) molecule (reprinted with permission from [28], copyright 1998, American Vacuum Society). (**c**) Tobacco mosaic virus (TMV) (reprinted from [88], with permission from Elsevier). (**d**) DNA molecules [30].

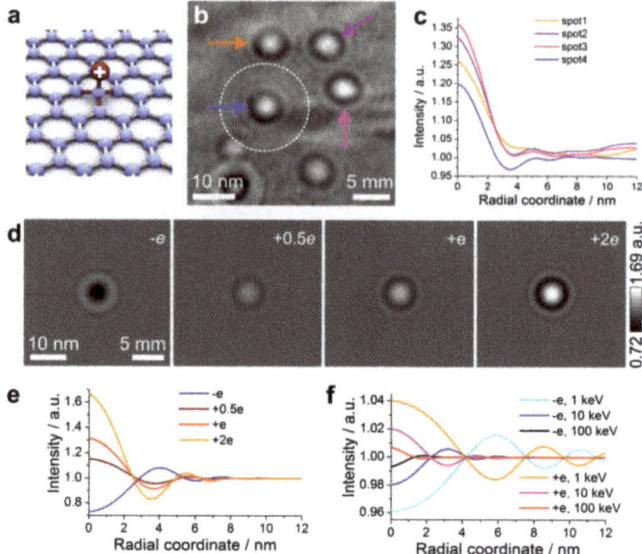

Figure 22. In-line electron holograms of charged adsorbates. (**a**) Schematic representation of a charged adsorbate on graphene. (**b**) Experimental hologram exhibiting bright spots; here, the electron energy is 30 eV, the source-to-sample distance is 82 nm and the source-to-screen distance is 47 mm. (**c**) Angular-averaged intensity profiles of the four bright spots marked in (**b**). (**d**) Simulated in-line holograms of a point charge, at four different values of charge, where the simulation parameters match those of the experimental hologram shown in (**b**). (**e**) Angular-averaged intensity profiles as a function of the radial coordinate, calculated from the simulated holograms shown in (**d**). (**f**) Angular-averaged intensity profiles as a function of the radial coordinate calculated from the simulated holograms at different high energies of probing electrons. The scale bars in (**b**,**d**) indicate the sizes in the object plane (left) and in the detector plane (right). Adapted with permission from [94], Copyright (2016) American Chemical Society.

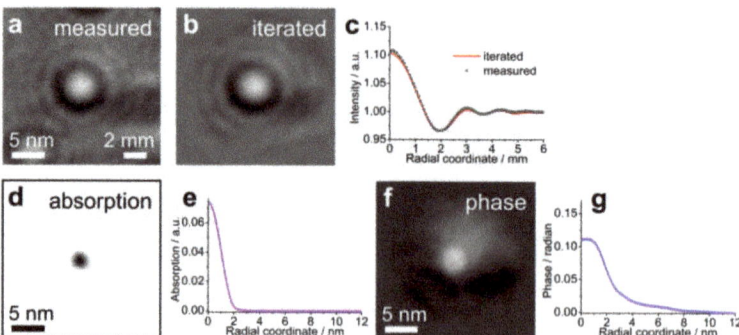

Figure 23. Iteratively reconstructed absorption and phase distribution of an individual charged impurity. (**a**) In-line hologram recorded with 30 eV electrons, exhibiting a bright spot. (**b**) Intensity distribution of the recovered wavefront obtained after 2000 iterations. (**c**) Angular-averaged radial profiles of the measured and iteratively recovered intensity distributions. (**d**,**f**) iteratively reconstructed absorption and phase distributions. (**e**,**g**) corresponding angular-averaged radial profiles. Reprinted from [74], with permission from Elsevier.

8.3. 3D Sample Reconstruction from Two or More In-Line Holograms

For a thin sample that can be described by a 2D transmission function, a single-shot in-line hologram is sufficient to reconstruct the absorption and phase distributions of the sample. However, realistic physical objects always have some finite thickness, and therefore are rather 3D than 2D samples. In optical holography, the complete reconstruction of a wavefront from a sequence of intensity measurements using an iterative procedure has been demonstrated in series of studies between 2003 and 2006 [95–99]. It has been recently demonstrated that 3D samples, including 3D phase objects, can be reconstructed from two or more holograms recorded at different z-distances from the sample [75], as illustrated in Figure 24. This reconstruction is performed by applying iterative phase retrieval only between the planes in which the intensity distributions were measured (H_1 and H_2 in Figure 24a), i.e., without involving any planes within the sample and hence with no need for constraints on the sample. The recovered complex-valued wavefront is then propagated back to the sample planes, and the 3D distribution of the sample is reconstructed (Figure 24b). It has been shown that in principle, as few as two holograms are sufficient to recover the entire wavefront diffracted by a 3D sample, and there is no restriction on the thickness of the sample or on the number of diffraction events within the sample. The sample does not need to be sparse, and a reference wave is not required. This method can be applied to 3D samples, such as a 3D distribution of particles, thick biological samples, and so on, including phase objects.

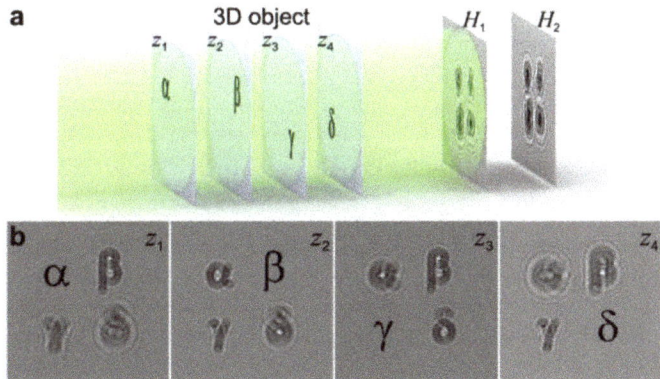

Figure 24. Reconstruction of 3D objects from two or more intensity measurements. (**a**) Experimental arrangement, in which the 3D sample is represented by a set of planes at different z-positions and two holograms are acquired at different distances from the sample, H_1 and H_2. (**b**) Reconstructed amplitude distributions at four planes within the 3D sample distribution. Adapted from [75].

9. Coherent Diffraction Imaging (CDI) with Electrons

9.1. CDI with High-Energy Electrons

CDI [100] is an imaging technique that is similar to diffraction in a crystal experiment, but involves imaging a single isolated object such as a macromolecule rather than a crystal [100–110]. In CDI, the structure of a sample is reconstructed from its diffraction pattern by applying an iterative phase retrieval algorithm [111–113]. To achieve this, the following requirements must be met: The object under study must be isolated; the size of the reconstructed field of view must exceed the size of the object by at least twice in each direction (oversampling condition) [113]; the incident wave must be a plane wave; and the imaging radiation must be coherent, although some attempts to employ partially coherent waves have been reported [114]. The power of CDI has been demonstrated by Zuo et al. who reconstructed the structure of a double-walled carbon nanotube (DWCNT) at atomic resolution from a diffraction pattern acquired using TEM with a nominal microscope point resolution of 2.2 Å for normal imaging at the Scherzer focus conditions [115], as shown in Figure 25. The results reported by Zuo et al. are often criticized, since the DWCNT had a finite size only in one dimension (x) and therefore the oversampling condition was not fulfilled for the other dimension (y), thus leading to an ambiguous reconstruction. However, it has recently been shown that for samples which can be described as a 1D chain of repeating units, or a 1D crystal, the average distribution of the repeating unit can be unambiguously reconstructed from the diffraction pattern of the sample, provided that the diffraction pattern is sufficiently oversampled [116]. Although the current study is limited to non-crystalline samples, we would like to add a notion that CDI of crystalline samples is highly challenging because of non-uniqueness of the reconstructed sample structure [117]. However, CDI can be successfully used for reconstruction of crystalline nano-particles when combined with other techniques [118,119].

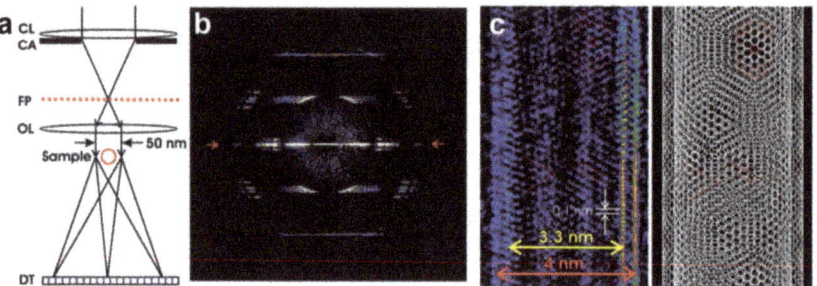

Figure 25. Coherent diffraction imaging of double-walled carbon nanotubes (DWCNTs) with high-energy electrons. (**a**) Schematic ray diagram of coherent nano-area electron diffraction. (**b**) Diffraction pattern of a DWCNT recorded with 200 keV electrons. (**c**) Section of the reconstructed DWCNT image at 1 Å resolution and (right) a structural model. Adapted from [115], reprinted with permission from AAAS.

9.2. CDI with Low-Energy Electrons

CDI with low-energy electrons has been demonstrated by Fink et al. in a dedicated low-energy electron microscope equipped with a microlens [120] to collimate the electron beam, as shown in Figure 26a. Low-energy electron diffraction patterns of individual macromolecules, such as carbon nanotubes [121,122] and graphene [9,123], were acquired. Diffraction patterns of individual stretched single-walled carbon nanotubes (SWCNTs) were acquired with 186 eV electrons at a resolution of 1.5 nm, as reported in reference [121], these results are shown in Figure 26b–d. Diffraction patterns of bundles of individual carbon nanotubes acquired with 145 eV electrons and reconstructed using a holographic CDI (HCDI) approach at a resolution of 0.7 nm were reported in reference [122], these are shown in Figure 26e–h.

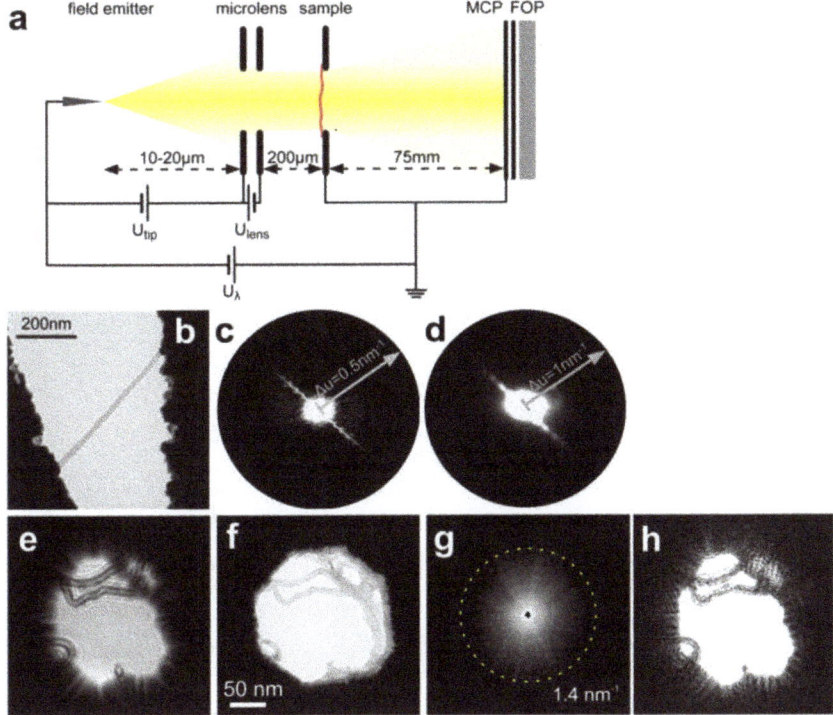

Figure 26. Coherent diffraction imaging (CDI) with low-energy electrons. (**a**) Experimental arrangement. (**b–d**) CDI of an individual single-walled carbon nanotube (SWCNT). (**b**) TEM image of the sample. (**c**) Fourier transform (FT) of the TEM image. (**d**) Diffraction pattern of SWCNTs recorded with 186 eV electrons. (**a–d**) reprinted from [121], with permission from Elsevier. (**e–h**) Holographic CDI (HCDI) reconstructions of a bundle of carbon nanotubes [122]. (**e**) In-line hologram recorded using electrons with kinetic energy 51 eV, source-to-sample distance 640 nm, and source-to-detector distance 68 mm. (**f**) TEM image recorded with 80 keV electrons. (**g**) Diffraction pattern recorded using electrons with kinetic energy 145 eV and source-to-detector distance 68 mm. The highest detected frequencies are indicated by the dashed circle, and the corresponding resolution is $R = \lambda/(2NA) = 7$ Å. (**h**) Reconstructed amplitude distribution of the sample using HCDI. (**e–h**) Adapted from [122].

10. Discussion

10.1. In-Line Holography (Defocus Imaging) vs. CDI

In this subsection, we compare the two imaging schemes of in-line holography and CDI, both schemes are shown in Figure 27. Each scheme has certain advantages and disadvantages, as summarized in Table 1.

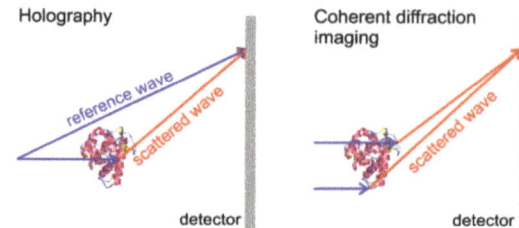

Figure 27. Schematic of (**a**) in-line holography and (**b**) coherent diffraction imaging (CDI).

Table 1. Advantages and disadvantages of in-line holography (defocus imaging) and CDI.

	In-Line Holography	CDI
Finding the sample in the microscope when imaging	Easy when imaging with widely expanded spherical wave (+)	Difficult when imaging with narrow collimated beam (−)
Phase information	Available from the recorded intensity (+)	Lost from the recorded intensity (−)
Reconstruction procedure	"One-step" reconstruction by calculating back-propagation integral (+)	Iterative reconstruction
Reconstructed information	z-information is available and a "three-dimensional" reconstruction is possible (+)	Reconstructed distribution is always a projection of the sample onto one plane (−)
Stability of the recorded image	Any lateral shift of the sample results in a lateral shift of the entire hologram (−)	Invariant to lateral shifts of the sample (+)
Resolution	Low resolution due to lateral and axial vibrations (−)	High resolution (+)

Radiation Dose

An important difference between in-line holography and CDI is the number of elastic scattering events (photons, electrons) required to obtain the sample reconstruction at a certain resolution. This difference is already evident from the principles of image formation in the two techniques: holography requires the object wave to be much weaker than the reference wave, while in CDI, only the object wave is measured, and it must therefore be sufficiently strong to be detected. In-line holography (or imaging in defocus) is often a suitable choice for imaging radiation-sensitive samples [124–127]. In the following, we provide a simple model for estimating and comparing the dose of radiation required in the two techniques to achieve a certain resolution. Here, we consider a phase object (phase 1 rad) of 10 nm in diameter probed with 200 keV electrons, although similar simulations and considerations can be done for electrons of different energy, or for photons.

In-line holography (Figure 28): In-line electron holograms (Figure 28a) were simulated with the following parameters: illuminated area is 100 nm × 100 nm, electron energy is 200 keV, incident wave is a spherical wave, and source-to-sample distance is 10 μm (the same hologram distributions can be obtained with a plane wave at a defocus distance of 10 μm [77]). An example of the simulated hologram is shown in Figure 28b. The radiation dose was changed from 1 to 100 particles per Å^2. For each radiation dose, a hologram was simulated and the hologram distribution was converged to integer numbers to mimic realistic detections of counts per pixel (cpp). The position of the highest detected interference fringes was then extracted, which defined the effective size of the hologram and the numerical aperture (NA). The resolution was calculated as $R = \lambda/(2NA)$ [122,128]. The resulting plot is shown in Figure 28c.

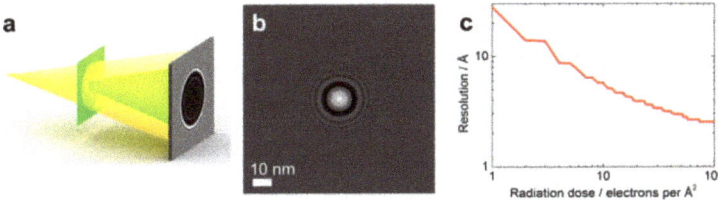

Figure 28. Radiation dose required to achieve a given resolution in in-line holography. (**a**) Experimental scheme for in-line holography. (**b**) Simulated in-line hologram of a round phase object of 10 nm in diameter. (**c**) Resolution as a function of radiation dose in in-line holography.

CDI (Figure 29): Diffraction patterns (Figure 29a) were simulated at radiation doses ranging from 1 to 10,000 electrons per Å^2. An example of simulated diffraction pattern is shown in Figure 29b. For the diffraction pattern simulated at a dose of 1 electron per Å^2, the angular-averaged radial profile is shown in Figure 28c; here the intensity rapidly decreases and reaches a threshold of 1 cpp at $k = 0.01$ Å^{-1} which corresponds to a resolution of about 10 nm. At each radiation dose, the simulated diffraction pattern distribution was converted to integer numbers (to mimic cpp), the angular-averaged radial profile was extracted, the position of k corresponding to a threshold of 1 cpp was determined and the resolution corresponding to that value of k was estimated. The resulting plot is shown in Figure 29d.

A comparison of the results shown in Figures 28c and 29d indicates that to achieve the same resolution, the radiation dose required by CDI is roughly a thousand times larger than that for in-line holography.

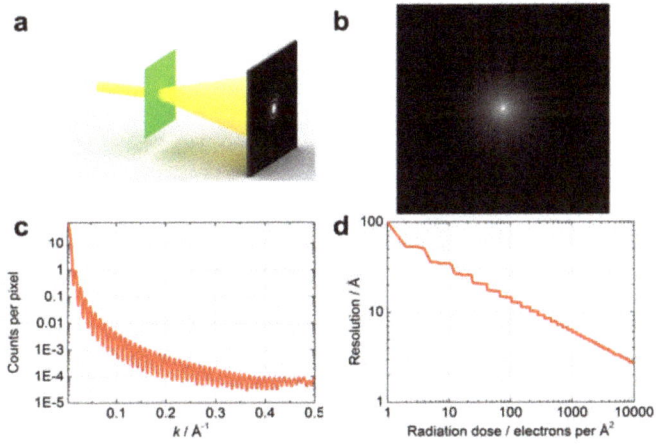

Figure 29. Radiation dose required to achieve a given resolution in CDI. (**a**) Experimental arrangement for CDI. (**b**) Simulated diffraction pattern of a round phase object of 10 nm in diameter. (**c**) Angular-averaged radial profile of a diffraction pattern simulated at a dose of 1 electron per Å^2. (**d**) Resolution as a function of radiation dose in CDI.

To provide an example of a more realistic sample, the diffraction pattern of a single lysozyme molecule was calculated at a radiation dose of 20 e/Å^2 with the multi-slice simulation protocol provided in Appendix A. The results are shown in Figure 30. The maximum intensity in the diffraction pattern is seen at the center and is 73 cpp; the intensity rapidly decreases and reaches a threshold of 1 cpp at $k = 0.035$ Å^{-1} which corresponds to a resolution of about 2.85 nm. However, the intensity distribution in the central region is usually not acquired in an experiment, because of intense direct beam. Similar simulated diffraction patterns of individual lysozyme molecules were presented by

Neutze et al. in a paper that addressed the possibility of single molecule diffraction with X-pulses from a free electron laser in a diffract-and-destroy experiment [129].

In summary, although CDI offers the theoretical possibility of recording high-resolution information, radiation damage limits the detected cpp in the diffraction pattern of a single molecule in practice, making it unsuitable to apply an iterative phase retrieval reconstruction routine to these diffraction patterns. To achieve a sufficiently strong signal for structure retrieval, other strategies such as averaging over thousands of diffraction patterns must be applied [130].

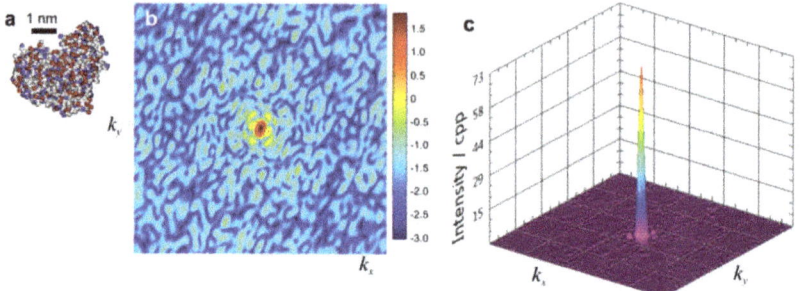

Figure 30. Simulated electron diffraction pattern of a single lysozyme molecule. (**a**) Structure of the lysozyme molecule. (**b**,**c**) simulated diffraction pattern with a radiation dose of 20 electrons per Å^2 in 2D and 3D representations, respectively; here k_x and k_y range from −0.5 Å^{-1} to 0.5 Å^{-1}, corresponding to a resolution at the rim of the diffraction pattern of 2 Å. The maximum of intensity is 73 cpp. Diffraction pattern (DP) in (**b**) is shown as $\log_{10}$(DP).

10.2. Low vs. High-Energy Electrons

Imaging with high-energy electrons is much more easily accessible, since high-energy electrons are employed in conventional TEMs and TEMs equipped with biprism(s) for off-axis holography experiments. Low-energy electron microscopes that operate in in-line holographic or point projection imaging regimes are self-built, and are not commercially available. High-energy electrons exhibit a relatively large IMFP, which allows for the imaging of materials tens of nanometers thick, while low-energy electrons can only image samples that are a few nanometers thick. Low-energy electrons are highly sensitive to potential distributions in an electron microscope, which causes artefactual deflection of the reference electron beam (the biprism effect) and complicates data analysis and evaluation of the sample structure [88,131]. To ensure an undisturbed reference wave, the sample needs to be placed onto an equipotential surface, for example graphene [92,93], in in-line low-energy electron holography. On the other hand, their high sensitivity to local potentials makes low-energy electrons the perfect type of radiation for studying 2D materials such as graphene and van der Waals structures [9,74,94,132]. Moreover, low-energy electrons can be employed for mapping unoccupied band structure of freestanding 2D materials, such as graphene by angle-resolved low-energy electron transmission measurements realized in in-line holography mode [133].

Funding: This research received no external funding.

Conflicts of Interest: The authors declare no conflict of interest.

Appendix A

The diffraction pattern of a single lysozyme molecule was calculated at a radiation dose of 20 e/Å^2 with the following multi-slice simulation protocol:

1. Lysozyme atomic coordinates were downloaded from PDB 253L [134], and hydrogen atoms were added using Chimera software.

2. The sequence of atoms was re-arranged in order of increasing z-coordinate, and atoms were numbered as a1, a2 etc.
3. An incident plane wave with unit amplitude was assumed, $u_1(x_1, y_1, z_1) = 1$.
4. The coordinates of the first atom a1 were read from the text file as $(x_1^{(1)}, y_1^{(1)}, z_1)$.
5. The transmission function in plane at z_1 was calculated as $t_1(x_1, y_1, z_1) = \exp[i\sigma v_z(x_1, y_1)]$, where σ is the interaction parameter at 200 keV and $v_z(x_1, y_1)$ is the projected potential of atom a1, calculated from the tabulated parameters corresponding to the chemical elements as described in reference [8].
6. The exit wave in the plane (x_1, y_1, z_1) was calculated as $u'_1(x_1, y_1, z_1) = u_1(x_1, y_1, z_1)t_1(x_1, y_1, z_1)$.
7. The z-coordinate of the next atom a2 was read as z_2, and the distance $\Delta z = z_2 - z_1$ was calculated.
8. The wave function $u'_1(x_1, y_1, z_1)$ was propagated for Δz using the angular spectrum method [77]. The resulting wavefront was $u_2(x_2, y_2, z_2)$.

The wave function was propagated through the sample, atom by atom, by repeating steps 4 to 8 until the electron wave had propagated through all the atoms. Finally, the diffraction pattern was calculated as square of the amplitude of the FT of the exit wave.

References

1. Broglie, L.D. Recherches sur la Théorie des Quanta. Ph.D. Thesis, Sorbonne Université, Paris, France, 1924.
2. Davisson, C.; Germer, L.H. The scattering of electrons by a single crystal of nickel. *Nature* **1927**, *119*, 558–560. [CrossRef]
3. NIST. *NIST Electron Elastic-Scattering Cross-Section Database*; National Institute of Standards and Technology: Gaithersburg, MA, USA, 2000.
4. Seah, M.P.; Dench, W.A. Quantitative electron spectroscopy of surfaces: A standard data base for electron inelastic mean free paths in solids. *Surf. Interface Anal.* **1979**, *1*, 2–11. [CrossRef]
5. Spence, J.C.H. STEM and shadow-imaging of biomolecules at 6 eV beam energy. *Micron* **1997**, *28*, 101–116. [CrossRef]
6. Ashley, J.C. Energy-loss rate and inelastic mean free-path of low-energy electrons and positrons in condensed matter. *J. Electron Spectrosc. Relat. Phenom.* **1990**, *50*, 323–334. [CrossRef]
7. Penn, D.R. Electron mean-free-path calculations using a model dielectric function. *Phys. Rev. B* **1987**, *35*, 482–486. [CrossRef]
8. Kirkland, E.J. *Advanced Computing in Electron Microscopy*; Springer: Berlin/Heidelberg, Germany, 2010.
9. Latychevskaia, T.; Escher, C.; Fink, H.-W. Moiré structures in twisted bilayer graphene studied by transmission electron microscopy. *Ultramicroscopy* **2019**, *197*, 46–52. [CrossRef]
10. Kasama, T.; Dunin-Borkowski, R.E.; Beleggia, M. Electron Holography of Magnetic Materials. In *Holography*; Ramirez, F.A.M., Ed.; IntechOpen: London, UK, 2011.
11. Goodman, J.W. *Introduction to Fourier Optics*, 3rd ed.; Roberts & Company Publishers: Greenwood Village, CO, USA, 2004.
12. Thompson, B.J.; Wolf, E. Two-beam interference with partially coherent light. *J. Opt. Soc. Am. A* **1957**, *47*, 895–902. [CrossRef]
13. Lichte, H.; Lehmann, M. Electron holography—Basics and applications. *Rep. Prog. Phys.* **2008**, *71*, 1–46. [CrossRef]
14. Van Cittert, P.H. Die wahrscheinliche Schwingungsverteilung in einer von einer Lichtquelle direkt oder mittels einer Linse beleuchteten Ebene. *Physica* **1934**, *1*, 201–210. [CrossRef]
15. Zernike, F. The concept of degree of coherence and its application to optical problems. *Physica* **1938**, *5*, 785–795. [CrossRef]
16. Goodman, J.W. *Statistical Optics*; Wiley Classics Library: Hoboken, NJ, USA, 1985.
17. Latychevskaia, T. Spatial coherence of electron beams from field emitters and its effect on the resolution of imaged objects. *Ultramicroscopy* **2017**, *175*, 121–129. [CrossRef] [PubMed]
18. Pozzi, G. Theoretical considerations on the spatial coherence in field-emission electron microscopes. *Optik* **1987**, *77*, 69–73.

19. Knoll, M.; Ruska, E. The electron microscope. *Z. Phys.* **1932**, *78*, 318–339. [CrossRef]
20. Iwanowski, D. Ueber die Mosaikkrankheit der Tabakspflanze. In *Bulletin Scientifique Publié Par l'Académie Impériale des Sciences de Saint-Pétersbourg/Nouvelle Serie III*; Imperial Academy of Arts: St.-Petersburg, Russia, 1892.
21. Beijerinck, M.W. *Ueber ein Contagium vivurn fluidum als Ursaehe der Fleekenkrankheit der Tabaksblätter. Verhandelingen der Koninklijke Akademie van Wetenschappen Te Amsterdam, Müller, Amsterdam*; KNAW: Amsterdam, The Netherlands, 1898.
22. Kausche, G.A.; Pfankuch, E.; Ruska, H. The visualisation of herbal viruses in surface microscopes. *Naturwissenschaften* **1939**, *27*, 292–299. [CrossRef]
23. Scherzer, O. Über einige Fehler von Elektronenlinsen. *Z. Phys.* **1936**, *101*, 593–603. [CrossRef]
24. Gabor, D. Improvements in and Relating to Microscopy. Patent GB685286, 17 December 1947.
25. Gabor, D. A new microscopic principle. *Nature* **1948**, *161*, 777–778. [CrossRef]
26. Gabor, D. Microscopy by reconstructed wave-fronts. *Proc. R. Soc. Lond. A* **1949**, *197*, 454–487. [CrossRef]
27. Morton, G.A.; Ramberg, E.G. Point projector electron microscope. *Phys. Rev.* **1939**, *56*, 705. [CrossRef]
28. Golzhauser, A.; Volkel, B.; Jager, B.; Zharnikov, M.; Kreuzer, H.J.; Grunze, M. Holographic imaging of macromolecules. *J. Vac. Sci. Technol. A* **1998**, *16*, 3025–3028. [CrossRef]
29. Golzhauser, A.; Volkel, B.; Grunze, M.; Kreuzer, H.J. Optimization of the low energy electron point source microscope: Imaging of macromolecules. *Micron* **2002**, *33*, 241–255. [CrossRef]
30. Eisele, A.; Voelkel, B.; Grunze, M.; Golzhauser, A. Nanometer resolution holography with the low energy electron point source microscope. *Z. Phys. Chem.* **2008**, *222*, 779–787. [CrossRef]
31. Beyer, A.; Golzhauser, A. Low energy electron point source microscopy: Beyond imaging. *J. Phys. Condes. Matter* **2010**, *22*, 343001. [CrossRef] [PubMed]
32. Beyer, A.; Weber, D.H.; Volkel, B.; Golzhauser, A. Characterization of nanowires with the low energy electron point source (LEEPS) microscope. *Phys. Status Solidi B* **2010**, *247*, 2550–2556. [CrossRef]
33. Vieker, H.; Beyer, A.; Blank, H.; Weber, D.H.; Gerthsen, D.; Golzhauser, A. Low energy electron point source microscopy of two-dimensional carbon nanostructures. *Z. Phys. Chem.* **2011**, *225*, 1433–1445. [CrossRef]
34. Hwang, I.-S.; Chang, C.-C.; Lu, C.-H.; Liu, S.-C.; Chang, Y.-C.; Lee, T.-K.; Jeng, H.-T.; Kuo, H.-S.; Lin, C.-Y.; Chang, C.-S.; et al. Investigation of single-walled carbon nanotubes with a low-energy electron point projection microscope. *New J. Phys.* **2013**, *15*, 043015. [CrossRef]
35. Muller, M.; Paarmann, A.; Ernstorfer, R. Visualization of photocurrents in nanoobjects by ultrafast low-energy electron point-projection imaging. In *Ultrafast Phenomena Xix*; Yamanouchi, I., Cundiff, S., De Vivie Riedle, R., Kuwata Gonokami, M., Di Mauro, L., Eds.; Springer-Verlag Berlin: Berlin, Germany, 2015; Volume 162, pp. 667–670.
36. Schmid, H.; Fink, H.-W. Combined electron and ion projection microscopy. *Appl. Surf. Sci.* **1993**, *67*, 436–443. [CrossRef]
37. Fink, H.-W.; Stocker, W.; Schmid, H. Holography with low-energy electrons. *Phys. Rev. Lett.* **1990**, *65*, 1204–1206. [CrossRef]
38. Fink, H.-W.; Schmid, H.; Ermantraut, E.; Schulz, T. Electron holography of individual DNA molecules. *J. Opt. Soc. Am. A* **1997**, *14*, 2168–2172. [CrossRef]
39. Mollenstedt, G.; Duker, H. Beobachtungen und Messungen an Biprisma-Interferenzen mit Elektronenwellen. *Z. Phys.* **1956**, *145*, 377–397. [CrossRef]
40. Mollenstedt, G.; Keller, M. Elektroneninterferometrisehe Messung des inneren Potentials. *Z. Phys.* **1957**, *148*, 34–37. [CrossRef]
41. Twitchett, A.C.; Dunin-Borkowski, R.E.; Midgley, P.A. Quantitative electron holography of biased semiconductor devices. *Phys. Rev. Lett.* **2002**, *88*, 238302. [CrossRef] [PubMed]
42. Twitchett, A.C.; Dunin-Borkowski, R.E.; Midgley, P.A. Comparison of off-axis and in-line electron holography as quantitative dopant-profiling techniques. *Philos. Mag.* **2006**, *86*, 5805–5823. [CrossRef]
43. Harscher, A.; Lichte, H. Determination of mean internal potential and mean free wavelength for inelastic scattering of vitrified iron by electron holography. *Eur. J. Cell Biol.* **1997**, *74*, 7.
44. Gatel, C.; Lubk, A.; Pozzi, G.; Snoeck, E.; Hytch, M. Counting elementary charges on nanoparticles by electron holography. *Phys. Rev. Lett.* **2013**, *111*, 025501. [CrossRef]

45. Vicarelli, L.; Migunov, V.; Malladi, S.K.; Zandbergen, H.W.; Dunin-Borkowski, R.E. Single electron precision in the measurement of charge distributions on electrically biased graphene nanotips using electron holography. *Nano Lett.* **2019**, *19*, 4091–4096. [CrossRef]
46. Aharonov, Y.; Bohm, D. Significance of electromagnetic potentials in the quantum theory. *Phys. Rev.* **1959**, *115*, 485–491. [CrossRef]
47. Chambers, R.G. Shift of an electron interfreence pattern by enclosed magnetic flux. *Phys. Rev. Lett.* **1960**, *5*, 3–5. [CrossRef]
48. Dunin-Borkowski, R.E.; McCartney, M.R.; Frankel, R.B.; Bazylinski, D.A.; Posfai, M.; Buseck, P.R. Magnetic microstructure of magnetotactic bacteria by electron holography. *Science* **1998**, *282*, 1868–1870. [CrossRef]
49. de Graef, M.; Nuhfer, N.T.; McCartney, M.R. Phase contrast of spherical magnetic particles. *J. Microsc. Oxf.* **1999**, *194*, 84–94. [CrossRef]
50. Thomas, J.M.; Simpson, E.T.; Kasama, T.; Dunin-Borkowski, R.E. Electron holography for the study of magnetic nanomaterials. *Acc. Chem. Res.* **2008**, *41*, 665–674. [CrossRef]
51. Midgley, P.A.; Dunin-Borkowski, R.E. Electron tomography and holography in materials science. *Nat. Mater.* **2009**, *8*, 271–280. [CrossRef] [PubMed]
52. Dunin-Borkowski, R.E.; Kasama, T.; Harrison, R.J. Chapter 5 Electron Holography of Nanostructured Materials. In *Nanocharacterisation (2)*; The Royal Society of Chemistry: London, UK, 2015; pp. 158–210. [CrossRef]
53. Latychevskaia, T.; Formanek, P.; Koch, C.T.; Lubk, A. Off-axis and inline electron holography: Experimental comparison. *Ultramicroscopy* **2010**, *110*, 472–482. [CrossRef]
54. Degiovanni, A.; Morin, R. Low Energy Electron Interferences Using A Biprism-Projection Microscope Combination. In *Electron Microscopy 1994, Vol 1—Interdisciplinary Developments and Tools*; Les Editions de Physique: Paris, France, 1994; pp. 331–332.
55. Morin, R. Point source physics: Application to electron projection microscopy and holography. *Microsc. Microanal. Microstruct.* **1994**, *5*, 501–508. [CrossRef]
56. Morin, P.; Pitaval, M.; Vicario, E. Low energy off-axis holography in electron microscopy. *Phys. Rev. Lett.* **1996**, *76*, 3979–3982. [CrossRef]
57. Morin, P. Computer simulation and object reconstruction in low-energy off-axis electron holography. *Ultramicroscopy* **1999**, *76*, 1–12. [CrossRef]
58. Degiovanni, A.; Bardon, J.; Georges, V.; Morin, R. Magnetic fields and fluxes probed by coherent low-energy electron beams. *Appl. Phys. Lett.* **2004**, *85*, 2938–2940. [CrossRef]
59. Lehmann, M.; Lichte, H. Tutorial on off-axis electron holography. *Microsc. Microanal.* **2002**, *8*, 447–466. [CrossRef]
60. Lichte, H. Electron interference: Mystery and reality. *Philos. Trans. R. Soc. A* **2002**, *360*, 897–920. [CrossRef]
61. Lichte, H.; Lehmann, H.W. Electron holography—A powerful tool for the analysis of nanostructures. *Adv. Imaging Elect. Phys.* **2002**, *123*, 225–255. [CrossRef]
62. Lichte, H.; Formanek, P.; Lenk, A.; Linck, M.; Matzeck, C.; Lehmann, M.; Simon, P. Electron holography: Applications to materials questions. *Annu. Rev. Mater. Res.* **2007**, *37*, 539–588. [CrossRef]
63. Simon, P.; Lichte, H.; Drechsel, J.; Formanek, P.; Graff, A.; Wahl, R.; Mertig, M.; Adhikari, R.; Michler, G.H. Electron holography of organic and biological materials. *Adv. Mater.* **2003**, *15*, 1475–1481. [CrossRef]
64. Simon, P.; Lichte, H.; Formanek, P.; Lehmann, M.; Huhle, R.; Carrillo-Cabrera, W.; Harscher, A.; Ehrlich, H. Electron holography of biological samples. *Micron* **2008**, *39*, 229–256. [CrossRef] [PubMed]
65. Lichte, H. Performance limits of electron holography. *Ultramicroscopy* **2008**, *108*, 256–262. [CrossRef] [PubMed]
66. Teague, M.R. Deterministic phase retrieval—A Green's function solution. *J. Opt. Soc. Am.* **1983**, *73*, 1434–1441. [CrossRef]
67. Schiske, P. Image reconstruction by means of focus series (Reprint of the original 1968 paper). *J. Microsc.* **2002**, *207*, 154. [CrossRef]
68. Coene, W.; Janssen, G.; Debeeck, M.O.; Vandyck, D. Phase retrieval through focus variation for ultra-resolution in field-emission transmission electron microscopy. *Phys. Rev. Lett.* **1992**, *69*, 3743–3746. [CrossRef]
69. Allen, L.J.; McBride, W.; O'Leary, N.L.; Oxley, M.P. Exit wave reconstruction at atomic resolution. *Ultramicroscopy* **2004**, *100*, 91–104. [CrossRef]

70. Allen, L.J.; Oxley, M.P. Phase retrieval from series of images obtained by defocus variation. *Opt. Commun.* **2001**, *199*, 65–75. [CrossRef]
71. Koch, C.T. A flux-preserving non-linear inline holography reconstruction algorithm for partially coherent electrons. *Ultramicroscopy* **2008**, *108*, 141–150. [CrossRef]
72. Latychevskaia, T.; Fink, H.-W. Solution to the twin image problem in holography. *Phys. Rev. Lett.* **2007**, *98*, 233901. [CrossRef]
73. Latychevskaia, T.; Fink, H.-W. Reconstruction of purely absorbing, absorbing and phase-shifting, and strong phase-shifting objects from their single-shot in-line holograms. *Appl. Opt.* **2015**, *54*, 3925–3932. [CrossRef]
74. Latychevskaia, T.; Wicki, F.; Escher, C.; Fink, H.-W. Imaging the potential distribution of individual charged impurities on graphene by low-energy electron holography. *Ultramicroscopy* **2017**, *182*, 276–282. [CrossRef] [PubMed]
75. Latychevskaia, T. Iterative phase retrieval for digital holography. *J. Opt. Soc. Am. A* **2019**, *36*, D31–D40. [CrossRef]
76. Latychevskaia, T. Reconstruction of missing information in diffraction patterns and holograms by iterative phase retrieval. *Opt. Commun.* **2019**, *452*, 56–67. [CrossRef]
77. Latychevskaia, T.; Fink, H.-W. Practical algorithms for simulation and reconstruction of digital in-line holograms. *Appl. Opt.* **2015**, *54*, 2424–2434. [CrossRef]
78. Latychevskaia, T.; Fink, H.-W. Simultaneous reconstruction of phase and amplitude contrast from a single holographic record. *Opt. Express* **2009**, *17*, 10697–10705. [CrossRef]
79. Gerchberg, R.W.; Saxton, W.O. Phase determination from image and diffraction plane pictures in electron microscope. *Optik* **1971**, *34*, 275–284.
80. Koren, G.; Joyeux, D.; Polack, F. Twin-image elimination in in-line holography of finite-support complex objects. *Opt. Lett.* **1991**, *16*, 1979–1981. [CrossRef]
81. Koren, G.; Polack, F.; Joyeux, D. Iterative algorithms for twin-image elimination in in-line holography using finite-support constraints. *J. Opt. Soc. Am. A* **1993**, *10*, 423–433. [CrossRef]
82. Latychevskaia, T.; Longchamp, J.-N.; Escher, C.; Fink, H.-W. Holography and coherent diffraction with low-energy electrons: A route towards structural biology at the single molecule level. *Ultramicroscopy* **2015**, *159*, 395–402. [CrossRef]
83. Spence, J.C.H.; Qian, W.; Melmed, A.J. Experimental low-voltage point-projection microscopy and its possibilities. *Ultramicroscopy* **1993**, *52*, 473–477. [CrossRef]
84. Spence, J.; Qian, W.; Zhang, X. Contrast and radiation-damage in point-projection electron imaging of purple membrane at 100-V. *Ultramicroscopy* **1994**, *55*, 19–23. [CrossRef]
85. Germann, M.; Latychevskaia, T.; Escher, C.; Fink, H.-W. Nondestructive imaging of individual biomolecules. *Phys. Rev. Lett.* **2010**, *104*, 095501. [CrossRef] [PubMed]
86. Latychevskaia, T.; Escher, C.; Andregg, W.; Andregg, M.; Fink, H.W. Direct visualization of charge transport in suspended (or free-standing) DNA strands by low-energy electron microscopy. *Sci. Rep.* **2019**, *9*, 8889. [CrossRef]
87. Latychevskaia, T.; Longchamp, J.-N.; Escher, C.; Fink, H.-W. Coherent Diffraction and Holographic Imaging of Individual Biomolecules Using Low-Energy Electrons. In *Advancing Methods for Biomolecular Crystallography*; Springer: Berlin/Heidelberg, Germany, 2013; pp. 331–342. [CrossRef]
88. Weierstall, U.; Spence, J.C.H.; Stevens, M.; Downing, K.H. Point-projection electron imaging of tobacco mosaic virus at 40 eV electron energy. *Micron* **1999**, *30*, 335–338. [CrossRef]
89. Longchamp, J.-N.; Latychevskaia, T.; Escher, C.; Fink, H.-W. Low-energy electron holographic imaging of individual tobacco mosaic virions. *Appl. Phys. Lett.* **2015**, *107*. [CrossRef]
90. Stevens, G.B.; Krüger, M.; Latychevskaia, T.; Lindner, P.; Plückthun, A.; Fink, H.-W. Individual filamentous phage imaged by electron holography. *Eur. Biophys. J.* **2011**, *40*, 1197–1201. [CrossRef]
91. Longchamp, J.-N.; Latychevskaia, T.; Escher, C.; Fink, H.-W. Non-destructive imaging of an individual protein. *Appl. Phys. Lett.* **2012**, *101*, 093701. [CrossRef]
92. Longchamp, J.-N.; Rauschenbach, S.; Abb, S.; Escher, C.; Latychevskaia, T.; Kern, K.; Fink, H.-W. Imaging proteins at the single-molecule level. *Proc. Natl. Acad. Sci. USA* **2017**, *114*, 1474–1479. [CrossRef]
93. Nair, R.R.; Blake, P.; Blake, J.R.; Zan, R.; Anissimova, S.; Bangert, U.; Golovanov, A.P.; Morozov, S.V.; Geim, A.K.; Novoselov, K.S.; et al. Graphene as a transparent conductive support for studying biological molecules by transmission electron microscopy. *Appl. Phys. Lett.* **2010**, *97*, 153102. [CrossRef]

94. Latychevskaia, T.; Wicki, F.; Longchamp, J.-N.; Escher, C.; Fink, H.-W. Direct observation of individual charges and their dynamics on graphene by low-energy electron holography. *Nano Lett.* **2016**, *16*, 5469–5474. [CrossRef] [PubMed]
95. Zhang, Y.; Pedrini, G.; Osten, W.; Tiziani, H.J. Whole optical wave field reconstruction from double or multi in-line holograms by phase retrieval algorithm. *Opt. Express* **2003**, *11*, 3234–3241. [CrossRef] [PubMed]
96. Pedrini, G.; Osten, W.; Zhang, Y. Wave-front reconstruction from a sequence of interferograms recorded at different planes. *Opt. Lett.* **2005**, *30*, 833–835. [CrossRef] [PubMed]
97. Almoro, P.; Pedrini, G.; Osten, W. Complete wavefront reconstruction using sequential intensity measurements of a volume speckle field. *Appl. Opt.* **2006**, *45*, 8596–8605. [CrossRef]
98. Li, Z.Y.; Li, L.; Qin, Y.; Li, G.B.; Wang, D.; Zhou, X. Resolution and quality enhancement in terahertz in-line holography by sub-pixel sampling with double-distance reconstruction. *Opt. Express* **2016**, *24*, 21134–21146. [CrossRef]
99. Guo, C.; Shen, C.; Li, Q.; Tan, J.B.; Liu, S.T.; Kan, X.C.; Liu, Z.J. A fast-converging iterative method based on weighted feedback for multi-distance phase retrieval. *Sci. Rep.* **2018**, *8*, 6436. [CrossRef]
100. Miao, J.W.; Charalambous, P.; Kirz, J.; Sayre, D. Extending the methodology of X-ray crystallography to allow imaging of micrometre-sized non-crystalline specimens. *Nature* **1999**, *400*, 342–344. [CrossRef]
101. Miao, J.W.; Hodgson, K.O.; Ishikawa, T.; Larabell, C.A.; LeGros, M.A.; Nishino, Y. Imaging whole Escherichia coli bacteria by using single-particle X-ray diffraction. *Proc. Natl. Acad. Sci. USA* **2003**, *100*, 110–112. [CrossRef]
102. Shapiro, D.; Thibault, P.; Beetz, T.; Elser, V.; Howells, M.; Jacobsen, C.; Kirz, J.; Lima, E.; Miao, H.; Neiman, A.M.; et al. Biological imaging by soft X-ray diffraction microscopy. *Proc. Natl. Acad. Sci. USA* **2005**, *102*, 15343–15346. [CrossRef]
103. Song, C.Y.; Jiang, H.D.; Mancuso, A.; Amirbekian, B.; Peng, L.; Sun, R.; Shah, S.S.; Zhou, Z.H.; Ishikawa, T.; Miao, J.W. Quantitative imaging of single, unstained viruses with coherent X-rays. *Phys. Rev. Lett.* **2008**, *101*, 158101. [CrossRef]
104. Williams, G.J.; Hanssen, E.; Peele, A.G.; Pfeifer, M.A.; Clark, J.; Abbey, B.; Cadenazzi, G.; de Jonge, M.D.; Vogt, S.; Tilley, L.; et al. High-resolution X-ray imaging of plasmodium falciparum-infected red blood cells. *Cytom. Part A* **2008**, *73A*, 949–957. [CrossRef] [PubMed]
105. Huang, X.; Nelson, J.; Kirz, J.; Lima, E.; Marchesini, S.; Miao, H.; Neiman, A.M.; Shapiro, D.; Steinbrener, J.; Stewart, A.; et al. Soft X-ray diffraction microscopy of a frozen hydrated yeast cell. *Phys. Rev. Lett.* **2009**, *103*, 198101. [CrossRef] [PubMed]
106. Nishino, Y.; Takahashi, Y.; Imamoto, N.; Ishikawa, T.; Maeshima, K. Three-dimensional visualization of a human chromosome using coherent X-ray diffraction. *Phys. Rev. Lett.* **2009**, *102*, 018101. [CrossRef] [PubMed]
107. Nelson, J.; Huang, X.J.; Steinbrener, J.; Shapiro, D.; Kirz, J.; Marchesini, S.; Neiman, A.M.; Turner, J.J.; Jacobsen, C. High-resolution X-ray diffraction microscopy of specifically labeled yeast cells. *Proc. Natl. Acad. Sci. USA* **2010**, *107*, 7235–7239. [CrossRef]
108. Wilke, R.N.; Priebe, M.; Bartels, M.; Giewekemeyer, K.; Diaz, A.; Karvinen, P.; Salditt, T. Hard X-ray imaging of bacterial cells: Nano-diffraction and ptychographic reconstruction. *Opt. Express* **2012**, *20*, 19232–19254. [CrossRef]
109. Seibert, M.M.; Ekeberg, T.; Maia, F.R.N.C.; Svenda, M.; Andreasson, J.; Jonsson, O.; Odic, D.; Iwan, B.; Rocker, A.; Westphal, D.; et al. Single mimivirus particles intercepted and imaged with an X-ray laser. *Nature* **2011**, *470*, 78–81. [CrossRef]
110. Ekeberg, T.; Svenda, M.; Abergel, C.; Maia, F.R.N.C.; Seltzer, V.; Claverie, J.-M.; Hantke, M.; Jönsson, O.; Nettelblad, C.; van der Schot, G.; et al. Three-dimensional reconstruction of the giant mimivirus particle with an X-ray free-electron laser. *Phys. Rev. Lett.* **2015**, *114*, 098102. [CrossRef]
111. Gerchberg, R.W.; Saxton, W.O. A practical algorithm for determination of phase from image and diffraction plane pictures. *Optik* **1972**, *35*, 237–246.
112. Fienup, J.R. Phase retrieval algorithms—A comparison. *Appl. Opt.* **1982**, *21*, 2758–2769. [CrossRef]
113. Latychevskaia, T. Iterative phase retrieval in coherent diffractive imaging: Practical issues. *Appl. Opt.* **2018**, *57*, 7187–7197. [CrossRef]

114. Whitehead, L.W.; Williams, G.J.; Quiney, H.M.; Vine, D.J.; Dilanian, R.A.; Flewett, S.; Nugent, K.A.; Peele, A.G.; Balaur, E.; McNulty, I. Diffractive imaging using partially coherent X-rays. *Phys. Rev. Lett.* **2009**, *103*, 243902. [CrossRef] [PubMed]
115. Zuo, J.M.; Vartanyants, I.; Gao, M.; Zhang, R.; Nagahara, L.A. Atomic resolution imaging of a carbon nanotube from diffraction intensities. *Science* **2003**, *300*, 1419–1421. [CrossRef] [PubMed]
116. Latychevskaia, T.; Fink, H.-W. Three-dimensional double helical DNA structure directly revealed from its X-ray fiber diffraction pattern by iterative phase retrieval. *Opt. Express* **2018**, *26*, 30991–31017. [CrossRef] [PubMed]
117. Wu, J.S.; Weierstall, U.; Spence, J.C.H. RETRACTED: Diffractive electron imaging of nanoparticles on a substrate (Retracted Article. See vol 5, pg 837, 2006). *Nat. Mater.* **2005**, *4*, 912–916. [CrossRef] [PubMed]
118. De Caro, L.; Carlino, E.; Caputo, G.; Cozzoli, P.D.; Giannini, C. Electron diffractive imaging of oxygen atoms in nanocrystals at sub-angstrom resolution. *Nat. Nanotechnol.* **2010**, *5*, 360–365. [CrossRef] [PubMed]
119. De Caro, L.; Carlino, E.; Vittoria, F.A.; Siliqi, D.; Giannini, C. Keyhole electron diffractive imaging (KEDI). *Acta Cryst. A* **2012**, *68*, 687–702. [CrossRef]
120. Steinwand, E.; Longchamp, J.-N.; Fink, H.-W. Fabrication and characterization of low aberration micrometer-sized electron lenses. *Ultramicroscopy* **2010**, *110*, 1148–1153. [CrossRef] [PubMed]
121. Steinwand, E.; Longchamp, J.-N.; Fink, H.-W. Coherent low-energy electron diffraction on individual nanometer sized objects. *Ultramicroscopy* **2011**, *111*, 282–284. [CrossRef]
122. Latychevskaia, T.; Longchamp, J.-N.; Fink, H.-W. When holography meets coherent diffraction imaging. *Opt. Express* **2012**, *20*, 28871–28892. [CrossRef]
123. Longchamp, J.-N.; Latychevskaia, T.; Escher, C.; Fink, H.-W. Graphene unit cell imaging by holographic coherent diffraction. *Phys. Rev. Lett.* **2013**, *110*, 255501. [CrossRef]
124. Adaniya, H.; Cheung, M.; Cassidy, C.; Yamashita, M.; Shintake, T. Development of a SEM-based low-energy in-line electron holography microscope for individual particle imaging. *Ultramicroscopy* **2018**, *188*, 31–40. [CrossRef] [PubMed]
125. Cheung, M.; Adaniya, H.; Cassidy, C.; Yamashita, M.; Shintake, T. Low-energy in-line electron holographic imaging of vitreous ice-embedded small biomolecules using a modified scanning electron microscope. *Ultramicroscopy* **2020**, *209*, 112883. [CrossRef] [PubMed]
126. Carlino, E. In-line holography in transmission electron microscopy for the atomic resolution imaging of single particle of radiation-sensitive matter. *Materials* **2020**, *13*, 1413. [CrossRef] [PubMed]
127. Zhou, L.; Song, J.; Kim, J.S.; Pei, X.; Huang, C.; Boyce, M.; Mendonça, L.; Clare, D.; Siebert, A.; Allen, C.S.; et al. Low-dose phase retrieval of biological specimens using cryo-electron ptychography. *Nat. Commun.* **2020**, *11*, 2773. [CrossRef] [PubMed]
128. Latychevskaia, T. Lateral and axial resolution criteria in incoherent and coherent optics and holography, near- and far-field regimes. *Appl. Opt.* **2019**, *58*, 3597–3603. [CrossRef] [PubMed]
129. Neutze, R.; Wouts, R.; van der Spoel, D.; Weckert, E.; Hajdu, J. Potential for biomolecular imaging with femtosecond X-ray pulses. *Nature* **2000**, *406*, 752–757. [CrossRef]
130. Branden, G.; Hammarin, G.; Harimoorthy, R.; Johansson, A.; Arnlund, D.; Malmerberg, E.; Barty, A.; Tangefjord, S.; Berntsen, P.; DePonte, D.P.; et al. Coherent diffractive imaging of microtubules using an X-ray laser. *Nat. Commun.* **2019**, *10*, 2589. [CrossRef]
131. Latychevskaia, T.; Longchamp, J.-N.; Escher, C.; Fink, H.-W. On artefact-free reconstruction of low-energy (30–250 eV) electron holograms. *Ultramicroscopy* **2014**, *145*, 22–27. [CrossRef]
132. Latychevskaia, T.; Hsu, W.-H.; Chang, W.-T.; Lin, C.-Y.; Hwang, I.-S. Three-dimensional surface topography of graphene by divergent beam electron diffraction. *Nat. Commun.* **2017**, *8*, 14440. [CrossRef]
133. Wicki, F.; Longchamp, J.-N.; Latychevskaia, T.; Escher, C.; Fink, H.-W. Mapping unoccupied electronic states of freestanding graphene by angle-resolved low-energy electron transmission. *Phys. Rev. B* **2016**, *94*, 075424. [CrossRef]
134. Shoichet, B.K.; Baase, W.A.; Kuroki, R.; Matthews, B.W. A relationship between protein stability and protein function. *Proc. Natl. Acad. Sci. USA* **1995**, *92*, 452–456. [CrossRef] [PubMed]

© 2020 by the author. Licensee MDPI, Basel, Switzerland. This article is an open access article distributed under the terms and conditions of the Creative Commons Attribution (CC BY) license (http://creativecommons.org/licenses/by/4.0/).

Article

Mapping the Magnetic Coupling of Self-Assembled Fe$_3$O$_4$ Nanocubes by Electron Holography

Lluís López-Conesa [1,2,3,*], Carlos Martínez-Boubeta [4,5], David Serantes [6], Sonia Estradé [1,2] and Francesca Peiró [1,2]

1. Laboratory of Electron Nanoscopies (LENS-MIND), Departament d'Enginyeria Electrònica i Biomèdica, Universitat de Barcelona, 08028 Barcelona, Spain; sestrade@ub.edu (S.E.); francesca.peiro@ub.edu (F.P.)
2. Institute of Nanoscience and Nanotechnology, Universtitat de Barcelona, (IN2UB), 08028 Barcelona, Spain
3. Centres Científics i Tecnològics de la Universitat de Barcelona (CCiTUB), 08028 Barcelona, Spain
4. Freelancer in O Con, 36950 Moaña, Spain; cboubeta@gmail.com
5. Ecoresources P.C., 54627 Thessaloniki, Greece
6. Instituto de Investigacións Tecnolóxicas and Departamento de Física Aplicada, Universidade de Santiago de Compostela, 15782 Santiago de Compostela, Spain; david.serantes@usc.es
* Correspondence: lluis.lopez.conesa@ub.edu

Abstract: The nanoscale magnetic configuration of self-assembled groups of magnetite 40 nm cubic nanoparticles has been investigated by means of electron holography in the transmission electron microscope (TEM). The arrangement of the cubes in the form of chains driven by the alignment of their dipoles of single nanocubes is assessed by the measured in-plane magnetic induction maps, in good agreement with theoretical calculations.

Keywords: electron holography; magnetic nanoparticles; magnetic hyperthermia

Citation: López-Conesa, L.; Martínez-Boubeta, C.; Serantes, D.; Estradé, S.; Peiró, F. Mapping the Magnetic Coupling of Self-Assembled Fe$_3$O$_4$ Nanocubes by Electron Holography. *Materials* **2021**, *14*, 774. https://doi.org/10.3390/ma14040774

Academic Editor: Marta Miola
Received: 30 December 2020
Accepted: 3 February 2021
Published: 6 February 2021

Publisher's Note: MDPI stays neutral with regard to jurisdictional claims in published maps and institutional affiliations.

Copyright: © 2021 by the authors. Licensee MDPI, Basel, Switzerland. This article is an open access article distributed under the terms and conditions of the Creative Commons Attribution (CC BY) license (https://creativecommons.org/licenses/by/4.0/).

1. Introduction

Magnetic hyperthermia has been the subject of intense research in recent years. Among the potential applications, it allows for a complementary approach to standard therapies for cancer treatment (for review, see e.g., [1]). This technique offers the advantage of delivering a highly localized damage via the targeting of tumor cells with magnetic nanoparticles. By exciting these nanoparticles with a radio-frequency signal, local heating of the surrounding area is achieved, with lower full-system toxicity than chemotherapy and without ionizing radiation affecting healthy tissue, as in the case of radiotherapy. However, in spite of having shown some promising results on palliative care, the high particle concentration required rises concerns about the toxicity and side effects of the treatment. Thus, improving efficiency by optimizing the magnetic response of nanoparticles is crucial in order to obtain therapeutic effects while keeping the number of nanoparticles as low as possible.

In this regard, performance is governed mainly by size distribution, saturation magnetization (M_S), and magnetic anisotropy (K) [2,3]. For a given excitation AC amplitude and frequency, these three are the parameters to tune in order to optimize the inductive specific absorption rate (SAR) of the system, usually reported in watts per gram [4]. To date, the highest reported SAR values correspond to metallic Fe nanocubes [5]. However, the low chemical stability of metallic nanoparticles under physiological conditions make the magnetically softer magnetite (Fe$_3$O$_4$) a much more promising candidate for applications in magnetic hyperthermia [6]. On the one hand, selecting Fe$_3$O$_4$ as the material of choice fixes a value for M_S. On the other hand, the particular application limits the range of particle sizes between the superparamagnetic limit ($\geq$15 nm) and the optimal size for internalization into mammalian cells ($\leq$50 nm) [7,8]. Thus, the remaining free parameters in order to optimize the heating response of the nanoparticles are the magnetic anisotropy (K) [9] and the volume fraction [4].

A way to increase magnetic anisotropy is by properly tuning the shape of the particles. Taking into account that a sphere has the minimum surface to volume ratio, cubic nanoparticles are already an improvement when compared to spherical ones because of their higher surface magnetic anisotropy. Another contribution to a larger surface anisotropy is the presence of well-defined atomic planes at the surfaces [10]: this is also in favor of the cubic shape, considering the most irregular crystal facets corresponding to a spherical nanoparticle.

An additional consequence of the cubic shape is an increased tendency of the magnetic nanoparticles to arrange in chains by sharing flat surfaces. The formation of ensembles of nanoparticles is also a way of engineering the magnetic response via the modification of the strength of the dipolar interaction between nanoparticles. Theoretical calculations for the hysteresis loops considering chains of Fe_3O_4 for different numbers of dipole-aligned nanocubes are reported in Boubeta et al. [11]. The simulations show an increasing area of the loop when increasing the number of aligned particles, therefore resulting in a potentiation of the heating efficiency. Furthermore, the thermal stability gained by creating arrays, also shown by simulations of magnetic response versus temperature, is an advantage when exploiting hysteresis losses. These results indicate a promising way to increase the hyperthermia performance by assembling cubic particles in elongated chains. On the heels of our previous article, here we use electron holography experiments to access and map the magnetic configuration of Fe_3O_4 cubic nanoparticles whose average diameter of 40 nm is expected to be close to the 180° domain wall width [12], thus may be promoting the presence of vortex pseudo-single-domain configurations [13,14].

2. Materials and Methods

Magnetite nanocube synthesis was performed following the one-pot and two-step procedure described previously [11]. Shortly, this requires the thermal decomposition of $Fe(acac)_3$ in boiling dibenzylether under argon atmosphere in the presence of decanoic acid. After cooling down, acetone was added to yield a precipitate, which was then separated by centrifugation. The supernatant was discarded and the particles were redispersed in chloroform. Samples for transmission electron microscopy (TEM) observation were prepared by dispersing a drop of the nanoparticle solution on a carbon-coated copper grid.

High resolution HRTEM experiments were carried out in a JEOL J2100 (Tokyo, Japan) located at CCiTUB. Electron holography experiments were carried out in the Hitachi I2TEM microscope (Tokyo, Japan) at CEMES-CNRS in Toulouse. The I2TEM is a modified Hitachi HF3300C TEM equipped with a 300 kV cold FEG, with an aberration corrector in the objective system and a 4k × 4k CCD camera. The I2TEM has an additional specimen holder port placed above the objective lens so that its magnetic field does not affect the specimen during the whole experiment. In this configuration, the aberration-corrected objective lens can be used as a Lorenz lens.

Micromagnetic simulations were performed with the OOMMF software package (version 1.0) [15], under the assumption that the nanocubes are perfectly cubic and identical. Each particle was discretized in 3D cells of 2 nm side, with a nonmagnetic intercube separation of 2 nm. We used bulk magnetic parameters for magnetite: M_S = 477 kA/m, cubic magnetocrystalline anisotropy K = −11 kJ/m^3, and exchange coupling constant of 1.0×10^{-11} J/m. The simulation procedure was to saturate the chains and let them relax to equilibrium at T = 0.

3. Results and Discussion

Our earlier studies [11] revealed a generalized self-assembly of Fe_3O_4 nanocubes in chain-like structures. Nanocubes are rather homogeneous in size, with ~40 nm lateral dimension. There was no apparent contrast variation within each nanoparticle, thus suggesting that particles were completely oxidized during synthesis. The magnetic properties of the particles are compatible with Fe_3O_4, with an incontrovertible evidence of Verwey transition around 120 K. HRTEM images confirmed monocrystalline Fe_3O_4 nanocubes indexed according to the inverse spinel structure of iron oxide.

Electron tomography [11] was used to reconstruct the 3D volume of a Fe_3O_4 nanoparticle chain. Results allowed accessing the shape of the chain in 3D and, at the same time, segmentation of the information down to single particle level. The cubic shape was confirmed by the 3D reconstruction, as well as cube alignment by sharing {100}-type flat faces. A separation in the order of ∼2 nm was found between adjacent cubes, corresponding to the organic ligand chains. At this surfactant layer thickness, van der Waals interaction between adjacent cubes is expected to be low [16], so the self-assembly could be ascribed to the magnetic dipole-dipole interaction.

Structural and morphological TEM characterization at the nanoscale, as well as macroscopic magnetic measurements, are in good agreement with the proposed model and the corresponding simulation reported previously [11]. However, this constitutes an indirect evidence of the magnetic coupling of the nanostructures. Direct evidence, namely real space imaging of the magnetic ordering down to single particle level, can be provided by electron holography [17–19].

In order to assess the magnetic state of the Fe_3O_4 ensembles, "up and down" electron holography experiments were carried out using two electrostatic biprisms. Which consists in acquiring two sets of holograms (sample and vacuum reference) corresponding to the two possible orientations of the TEM specimen. This requires taking the sample out of the microscope and flipping it between the two acquisitions. A hologram is formed by the superposition of two electron beams on the detector: one beam has travelled through the specimen and the other one has travelled through vacuum. The superposition of the two beams is obtained using an electrostatic biprism (in our setup, the lower one), as depicted in Figure 1. The resulting hologram contains interference fringes due to the phase shift caused by the specimen on the electron beam that travelled through it.

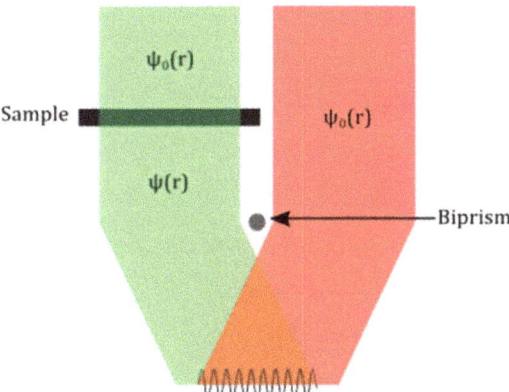

Figure 1. Off axis electron holography basic diagram. The electron wave resulting from the interaction of the electron beam with the sample, and a reference electron wave from the electron beam travelling along the vacuum, are made to interfere using an electrostatic biprism. The resulting interference fringe pattern is studied.

Figure 2a,c show the two flip-related holograms for an ensemble of nanocubes. The use of two electrostatic biprisms allows decoupling two important parameters: the width of the superposition region and the interference fringes spacing [20]. When working in a single biprism configuration, the applied voltage defines both parameters, so that a balance needs to be found. The use of two biprisms allows controlling them separately by defining different voltages for each one of them. An additional advantage of this configuration is the elimination of Fresnel interference fringes in the holograms when the lower biprism is in the region shaded by the upper one. This can be seen in Figure 2b, where only the centerband and the sidebands are present in the Fourier transform. This results in a higher fringe contrast, which is a key parameter limiting the magnetic signal resolution. The obtained small fringe spacing and high contrast in the recorded holograms is illustrated in Figure 2d.

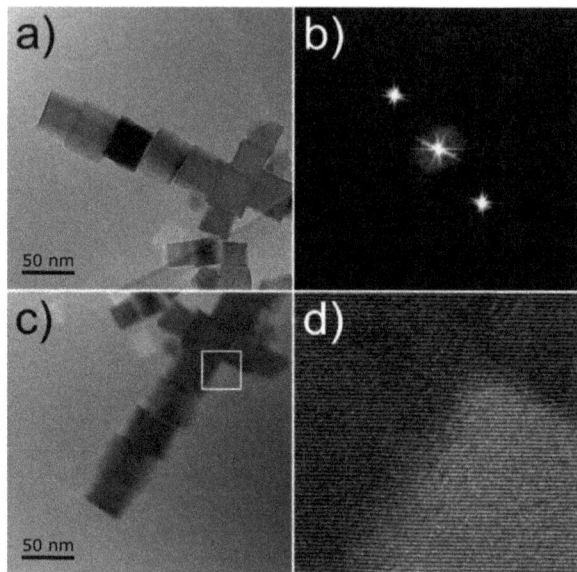

Figure 2. (a) Hologram covering a nanoparticle ensemble formed by two crossing chains. (b) FFT of the hologram showing the center band and sidebands. No spots corresponding to Fresnel fringes are visible due to the use of a two biprism configuration. (c) Hologram of the same ensemble obtained after mechanical flipping of the sample. (d) Detail of the interference fringes from the region highlighted in (c).

After subtracting the constant phase term corresponding to the vacuum reference holograms for both the up and down configurations (not shown here), and correcting the images for the mechanical flip process, a mask is set on one of the sidebands, and the corresponding amplitude and phase are calculated. The obtained phase shift maps for the up and down holograms of the ensemble under study are shown in Figure 3a,c, respectively. Considering the experimental setup, the only actual contributions to the phase shift (φ) are the electrostatic and the magnetic phases. Each one of the phase maps will have contributions from both electrostatic (φ_E) and magnetic (φ_M) components

$$\varphi_{up} = \varphi_{E,up} + \varphi_{M,up} \tag{1}$$

$$\varphi_{down} = \varphi_{E,down} + \varphi_{M,down} \tag{2}$$

Given the flipping process between the two acquisitions and the nature of the electrostatic and magnetic fields, the phase shifts resulting from the two holograms will satisfy the following relationship

$$\varphi_{E,up} = \varphi_{E,down} \tag{3}$$

$$\varphi_{M,up} = -\varphi_{M,down} \tag{4}$$

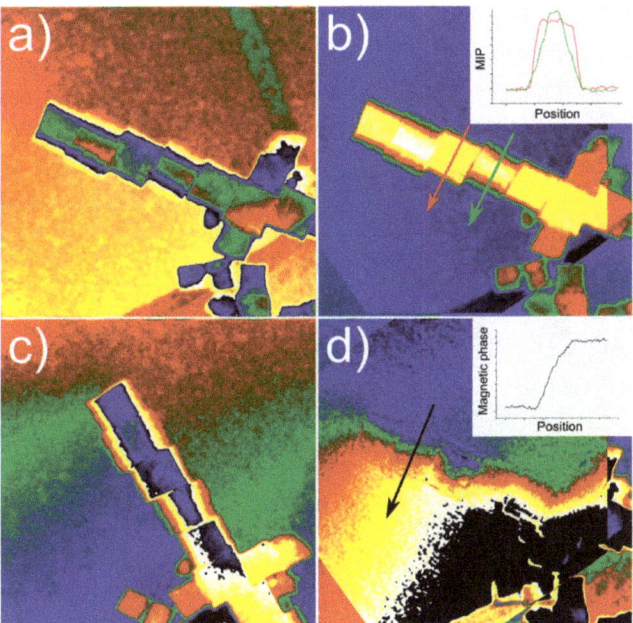

Figure 3. (**a,c**) Phase shift maps corresponding to the "up" and "down" holograms, respectively. The "down" image needs to be flipped so that it can be aligned with the "top" image. (**b**) Phase sum image, corresponding to the mean inner potential (MIP). (**d**) Phase difference image, corresponding to the magnetic phase shift. The magnetic phase difference across the object, in the direction of the black arrow and shown in the inset, indicates its magnetic nature.

So, after careful alignment of the phase shift maps, simple phase operations allow separating the magnetic phase

$$\varphi_M = \frac{\varphi_{up} - \varphi_{down}}{2} \quad (5)$$

from the electrostatic phase corresponding to the mean inner potential (MIP) of the sample.

$$\varphi_E = \frac{\varphi_{up} + \varphi_{down}}{2} \quad (6)$$

The resulting phase sum and difference maps are shown in Figure 3b,d, respectively. The dependence of the MIP is on the electric charge distribution and sample thickness so, considering a homogeneous material, an intensity profile across the sample can provide information on the third dimension. The MIP intensity profiles show sharper edges for the cube presenting a stronger diffraction contrast, as could be expected from a cube lying flat on one face and therefore closer to the zone axis. The phase difference map corresponding to the magnetic signal shows a phase shift with a frontier laying along the direction of the nanocube chain. This magnetic phase difference, clearly shown in the intensity profile in the inset, is a clear signature of the magnetic behavior of the nanocubes.

From the obtained magnetic phase φ_M, the magnetic induction map in the specimen plane (perpendicular to the beam direction, $B^p(x,y)$) can be calculated as its gradient

$$\vec{\nabla}\varphi_M(x,y) = \frac{e}{\hbar}\left[B_y^p(x,y) - B_x^p(x,y)\right] \quad (7)$$

A different way to visualize the magnetic coupling along the chain is by representing the magnetic phase shift as contour maps according to the expression $\cos(n\varphi_M)$ for n = 1, 2, ... The resulting contours represent the change in magnetic phase and, thus, constitute a map of the in-plane magnetic flux lines. The induction vector map and contour map for the ensemble under study are shown in Figure 4b,c, next to the inverse fast Fourier transform (IFFT) of the hologram centerband as a geometrical reference (Figure 4a). The magnetic signal is somewhat distorted in the central nanocube showing a stronger diffraction contrast due to its crystal orientation, as mentioned before for the MIP map. Diffraction contrast decreases the interference fringes contrast, thus making the detection of the magnetic signal difficult. In this example, the ensemble is formed by two crossing chains: a long chain with N = 6 along the vertical direction and a shorter horizontal one with N = 3. Magnetic flux lines follow the alignment of the chains and rotate ~55° in the "node" nanocube where the two chains intersect at a right angle.

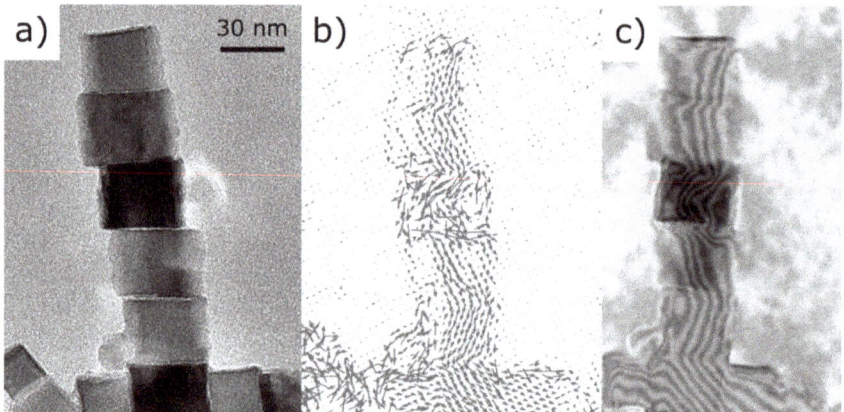

Figure 4. (a) Reference images. (b) In-plane induction map. (c) Magnetic phase signal visualized as a $\cos(n\varphi_M)$ contour map superimposed to the amplitude image. Arrows and contours correspond to magnetic flux lines showing the magnetic coupling of the cubes. Stray field lines are visible at the tip of the chain.

One wonders whether such a peculiar magnetic configuration could be reproduced by micromagnetic calculations. In a naive picture, we can see the two chains depicted in Figure 4 as a T-shaped structure. As the sample has never been exposed to any magnetic field, the measured configurations should correspond to virgin remnant states. The results are shown in Figure 5 and correspond rather nicely to the experimental ones. On the one hand, the elongated structure introduces a uniaxial anisotropy of magnetostatic origin and defines the easy axis for the magnetization. On the other hand, the surfaces of the nanocubes correspond to [100] planes and as the magnetization attempts to flip between <111> easy crystallographic directions, the spins curling in the junction must have opposite helicities [17], and form an angle of $\theta = \cos^{-1}(1/\sqrt{3}) \sim 55°$.

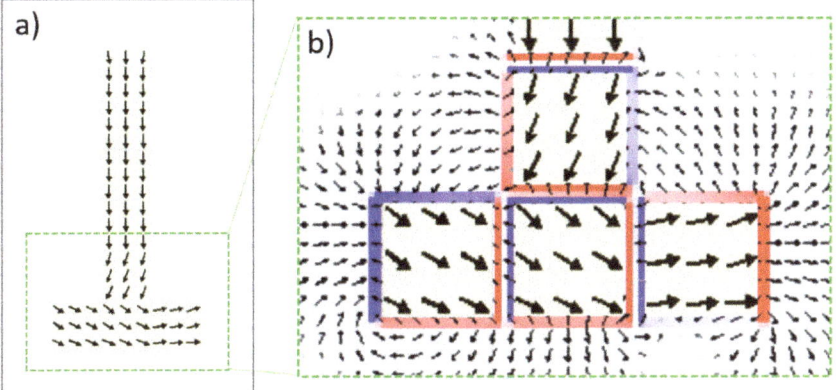

Figure 5. (**a**) Micromagnetic configuration at remanence of the magnetization of two crossing chains as those depicted in Figure 4. To ease the observation, each arrows stand for average magnetization over 7 × 7 × 7 unit cells. (**b**) Augmented view of the crossing-chains area, superimposed with the stray field configuration depicted by the small arrows external to the magnetic cubes (light yellow regions); for illustrative purposes the stray field arrows are averaged over 3 × 3 × 3 unit cells. The red-blue colors indicate the divergence of the stray fields.

An analogous processing was carried out for holograms from different assemblies and the resulting induction vector maps and contour maps of the magnetic phase shift are shown in Figure 6. Both of them present a cooperative organization governed by the dipole–dipole interaction, despite their stronger spatial deviation from a perfectly aligned assembly. This is probably because they contain a bigger proportion of crystals of different sizes. Long reaching stray field lines are visible, particularly at the tips, both on simulated and experimental mappings, but close outside the field of view. Flux lines forming concentric circles can also be seen in Figure 6e,f. The contrast spot observed at the center of that nanocube corresponds to the turn out-of-plane magnetization [21], which leads to a drastic reduction of the dipolar energy.

We will end by making at least a brief reference to such vortex configurations. Figure 7 shows the size dependence of the spontaneous magnetization of a cubic magnetite nanoparticle. With increasing particle sizes beyond ~50 nm, the magnetization of a single domain vanishes indicating the 3D vortex flux closure structure. Additionally, for exploratory purposes we included (not-shown) an iron oxide outer layer of thickness 0–2 nm with the bulk maghemite magnetic parameters. This thin shell layer, however, does not seem to change the simulated magnetic configurations of the Fe_3O_4 nanocubes.

Similar calculations have been performed in the past, especially by Butler and Banerjee [12]. They found that stable single-domain cubic magnetite nanoparticles at 290 K exist in the transition region 40–76 nm imposed by the superparamagnetic limit and the cost of introducing domain walls. Accordingly, in another study Usov et al. [22] estimated it in about 56 nm in lateral size. Therefore, the overall agreement is reasonable considering the experimental errors and the zero temperature simulations. Moreover, a vortex-like state such the one depicted in Figure 6, which is now perpendicular to the chain axis, may also depend sensitively on the particular arrangement of the surrounding assembles [23], the explanation of which is beyond the scope of this paper.

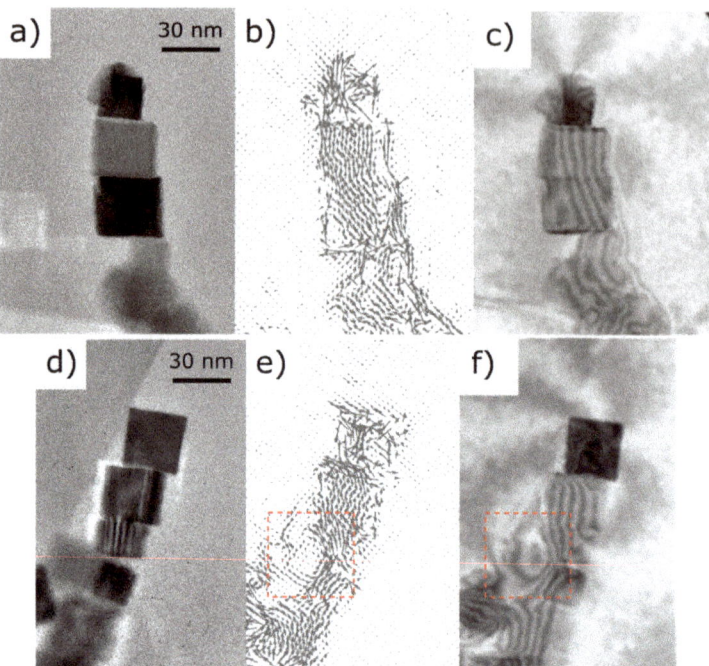

Figure 6. (**a,d**) Reference images. (**b,e**) In-plane induction maps. (**c,f**) Magnetic phase signal visualized as a cos(nφ_M) contour map superimposed to the amplitude image. Arrows and contours correspond to magnetic flux lines showing the magnetic coupling of the cubes. Stray field lines are visible at the tip of the chains. A vortex spinning perpendicular to the chain axis is also visible in (**e,f**), highlighted by red squares. The two last crystals on the upper side of the chains show a complex magnetic state. Interactions between neighboring nanocubes induce a bending of the magnetic induction.

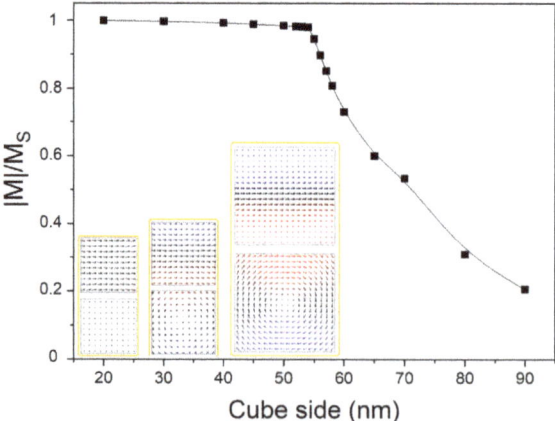

Figure 7. Size dependence of normalized magnetization. The snapshots show the remanent magnetization configurations, taken along two orthogonal directions, for exemplary dimensions (50, 58, and 90 nm). For clarity purposes the arrows representing the magnetization are average over 10 × 10 × 10 basic unit cells. All the particles are drawn at the same scale.

4. Conclusions

We were able to map the magnetic configuration of ensembles of Fe_3O_4 nanocubes, approximately 40 nm in size, by means of electron holography. The self-assembly of the nanocubes in the form of chains was confirmed to be driven both by the shapes of the blocks and by the dipole–dipole interaction. Furthermore, a very good agreement between simulated and experimental phase shift maps is obtained.

In this regard, our former work unambiguously demonstrate the important role of chain alignment on the area of hysteresis loop (and therefore of the SAR) [24]. Consequently, disorientation of the assembly and deviations from the homogeneous flux distribution (as the ones reported in Figure 6) would lead to a considerable decrease in the heating efficiency and can most probably explain the smaller SAR values for the 40 nm sample compared to the 20 nm case [11].

It is our view that our findings contribute to the knowledge on the complexity of the magnetic structure in applications as diverse as non-volatile storage devices and cancer therapies, which calls for further studies.

Author Contributions: Conceptualization, L.L.-C., S.E. and F.P.; formal analysis, L.L.-C., C.M.-B. and D.S.; funding acquisition, C.M.-B. and F.P.; investigation, L.L.-C.; methodology, L.L.-C and C.M.-B.; project administration, F.P.; resources, C.M.-B. and F.P.; supervision, S.E. and F.P.; visualization, L.L.-C. and D.S.; writing—original draft, L.L.-C.; writing—review & editing, C.M.-B., D.S., S.E. and F.P. All authors have read and agreed to the published version of the manuscript.

Funding: This research received funding from the Xunta de Galicia, Program for Development of a Strategic Grouping in Materials (AeMAT, Grant No. ED431E2018/08), the Agencia Estatal de Investigación (project PID2019-109514RJ-100), the Spanish Ministry of Economy and Competitiveness, grant number MAT2010-16407, and from the European Union Seventh Framework Programme under Grant Agreement 312483-ESTEEM2 (Integrated Infrastructure Initiative–I3).

Acknowledgments: We thank all our colleagues from ref. [11], and also Catalina Coll, who provided insight and expertise that greatly assisted the research. We acknowledge the use of computational facilities at the Centro de Supercomputación de Galicia (CESGA).

Conflicts of Interest: The authors declare no conflict of interest.

References

1. Kumar, C.S.S.R.; Mohammad, F. Magnetic nanomaterials for hyperthermia-based therapy and controlled drug delivery. *Adv. Drug Deliv. Rev.* **2011**, *63*, 789–808. [CrossRef] [PubMed]
2. Butter, K.; Bomans, P.H.H.; Frederik, P.M.; Vroege, G.J.; Philipse, A.P. Direct observation of dipolar chains in iron ferrofluids by cryogenic electron microscopy. *Nat. Mater.* **2003**, *2*, 88–91. [CrossRef] [PubMed]
3. Majetich, S.A.; Wen, T.; Booth, R.A. Functional Magnetic Nanoparticle Assemblies: Formation, Collective Behavior, and Future Directions. *ACS Nano* **2011**, *5*, 6081–6084. [CrossRef]
4. Conde-Leboran, I.; Baldomir, D.; Martinez-Boubeta, C.; Chubykalo-Fesenko, O.; Del Puerto Morales, M.; Salas, G.; Cabrera, D.; Camarero, J.; Teran, F.J.; Serantes, D. A Single Picture Explains Diversity of Hyperthermia Response of Magnetic Nanoparticles. *J. Phys. Chem. C* **2015**, *119*, 15698–15706. [CrossRef]
5. Mehdaoui, B.; Meffre, A.; Carrey, J.; Lachaize, S.; Lacroix, L.-M.; Gougeon, M.; Chaudret, B.; Respaud, M. Optimal Size of Nanoparticles for Magnetic Hyperthermia: A Combined Theoretical and Experimental Study. *Adv. Funct. Mater.* **2011**, *21*, 4573–4581. [CrossRef]
6. Soetaert, F.; Korangath, P.; Serantes, D.; Fiering, S.; Ivkov, R. Cancer therapy with iron oxide nanoparticles: Agents of thermal and immune therapies. *Adv. Drug Deliv. Rev.* **2020**, *163–164*, 65–83. [CrossRef]
7. Fortin, J.-P.; Gazeau, F.; Wilhelm, C. Intracellular heating of living cells through Néel relaxation of magnetic nanoparticles. *Eur. Biophys. J.* **2008**, *37*, 223–228. [CrossRef]
8. Zhang, S.; Li, J.; Lykotrafitis, G.; Bao, G.; Suresh, S. Size-Dependent Endocytosis of Nanoparticles. *Adv. Mater.* **2009**, *21*, 419–424. [CrossRef]
9. Lacroix, L.-M.; Malaki, R.B.; Carrey, J.; Lachaize, S.; Respaud, M.; Goya, G.F.; Chaudret, B. Magnetic hyperthermia in single-domain monodisperse FeCo nanoparticles: Evidences for Stoner-Wohlfarth behavior and large losses. *J. Appl. Phys.* **2009**, *105*, 023911. [CrossRef]
10. Skomski, R.; Wei, X.-H.; Sellmyer, D.J.; Skomski, R.; Sellmyer, D.J. Magnetization Reversal in Cubic Nanoparticles With Uniaxial Surface Anisotropy. *IEEE Trans. Magn.* **2007**, *43*. [CrossRef]

11. Martinez-Boubeta, C.; Simeonidis, K.; Makridis, A.; Angelakeris, M.; Iglesias, O.; Guardia, P.; Cabot, A.; Yedra, L.; Estradé, S.; Peiró, F.; et al. Learning from nature to improve the heat generation of iron-oxide nanoparticles for magnetic hyperthermia applications. *Sci. Rep.* **2013**, *3*, 1652. [CrossRef]
12. Butler, R.F.; Banerjee, S.K. Theoretical single-domain grain size range in magnetite and titanomagnetite. *J. Geophys. Res.* **1975**, *80*, 4049–4058. [CrossRef]
13. Simeonidis, K.; Martinez-Boubeta, C.; Serantes, D.; Ruta, S.; Chubykalo-Fesenko, O.; Chantrell, R.; Oró-Solé, J.; Balcells, L.; Kamzin, A.S.; Nazipov, R.A.; et al. Controlling Magnetization Reversal and Hyperthermia Efficiency in Core-Shell Iron-Iron Oxide Magnetic Nanoparticles by Tuning the Interphase Coupling. *ACS Appl. Nano Mater.* **2020**, *3*, 4465–4476. [CrossRef]
14. Bonilla, F.J.; Lacroix, L.M.; Blon, T. Magnetic ground states in nanocuboids of cubic magnetocrystalline anisotropy. *J. Magn. Magn. Mater.* **2017**, *428*, 394–400. [CrossRef]
15. OOMMF Project at NIST. Available online: https://math.nist.gov/oommf/ (accessed on 26 December 2020).
16. Liao, H.-G.; Cui, L.; Whitelam, S.; Zheng, H. Real-Time Imaging of Pt3Fe Nanorod Growth in Solution. *Science* **2012**, *336*, 1011–1014. [CrossRef] [PubMed]
17. Snoeck, E.; Gatel, C.; Lacroix, L.M.; Blon, T.; Lachaize, S.; Carrey, J.; Respaud, M.; Chaudret, B. Magnetic configurations of 30 nm iron nanocubes studied by electron holography. *Nano Lett.* **2008**, *8*, 4293–4298. [CrossRef]
18. Shindo, D.; Murakami, Y. Electron holography of magnetic materials. *J. Phys. D Appl. Phys.* **2008**, *41*, 183002. [CrossRef]
19. Reichel, V.; Kovács, A.; Kumari, M.; Bereczk-Tompa, É.; Schneck, E.; Diehle, P.; Pósfai, M.; Hirt, A.M.; Duchamp, M.; Dunin-Borkowski, R.E.; et al. Single crystalline superstructured stable single domain magnetite nanoparticles. *Sci. Rep.* **2017**, *7*, 45484. [CrossRef]
20. Harada, K.; Tonomura, A.; Togawa, Y.; Akashi, T.; Matsuda, T. Double-biprism electron interferometry. *Appl. Phys. Lett.* **2004**, *84*, 3229–3231. [CrossRef]
21. Shinjo, T.; Okuno, T.; Hassdorf, R.; Shigeto, K.; Ono, T. Magnetic vortex core observation in circular dots of permalloy. *Science* **2000**, *289*, 930–932. [CrossRef]
22. Usov, N.A.; Nesmeyanov, M.S.; Tarasov, V.P. Magnetic Vortices as Efficient Nano Heaters in Magnetic Nanoparticle Hyperthermia. *Sci. Rep.* **2018**, *8*, 1–9. [CrossRef]
23. Dunin-Borkowski, R.E.; Kasama, T.; Wei, A.; Tripp, S.L.; Hÿtch, M.J.; Snoeck, E.; Harrison, R.J.; Putnis, A. Off-axis electron holography of magnetic nanowires and chains, rings, and planar arrays of magnetic nanoparticles. *Microsc. Res. Tech.* **2004**, *64*, 390–402. [CrossRef]
24. Serantes, D.; Simeonidis, K.; Angelakeris, M.; Chubykalo-Fesenko, O.; Marciello, M.; Del Puerto Morales, M.; Baldomir, D.; Martinez-Boubeta, C. Multiplying magnetic hyperthermia response by nanoparticle assembling. *J. Phys. Chem. C* **2014**, *118*, 5927–5934. [CrossRef]

Article

Study of the Microstructure of Amorphous Silica Nanostructures Using High-Resolution Electron Microscopy, Electron Energy Loss Spectroscopy, X-ray Powder Diffraction, and Electron Pair Distribution Function

Lahcen Khouchaf [1], Khalid Boulahya [2], Partha Pratim Das [3,4,*], Stavros Nicolopoulos [4,*], Viktória Kovács Kis [5] and János L. Lábár [5]

1. École Nationale Supérieure des Mines-Télécom de Lille-Douai Lille Douai, Lille Université, CEDEX, 59653 Villeneuve D'Ascq, France; lahcen.khouchaf@imt-lille-douai.fr
2. Departamento de Química Inorgánica, Facultad de Químicas, Universidad Complutense, 28040 Madrid, Spain; khalid@quim.ucm.es
3. Electron Crystallography Solutions SL, Calle Orense 8, 28020 Madrid, Spain
4. NanoMEGAS SPRL, Blvd Edmond Machtens 79, B-1080 Brussels, Belgium
5. Institute of Technical Physics and Materials Science, Centre for Energy Research, 1121 Budapest, Hungary; kis.viktoria@energia.mta.hu (V.K.K.); labar.janos@energia.mta.hu (J.L.L.)
* Correspondence: partha@ecrystsolutions.com (P.P.D.); info@nanomegas.com (S.N.)

Received: 29 July 2020; Accepted: 27 September 2020; Published: 1 October 2020

Abstract: Silica has many industrial (i.e., glass formers) and scientific applications. The understanding and prediction of the interesting properties of such materials are dependent on the knowledge of detailed atomic structures. In this work, amorphous silica subjected to an accelerated alkali silica reaction (ASR) was recorded at different time intervals so as to follow the evolution of the structure by means of high-resolution transmission electron microscopy (HRTEM), electron energy loss spectroscopy (EELS), and electron pair distribution function (e-PDF), combined with X-ray powder diffraction (XRPD). An increase in the size of the amorphous silica nanostructures and nanopores was observed by HRTEM, which was accompanied by the possible formation of Si–OH surface species. All of the studied samples were found to be amorphous, as observed by HRTEM, a fact that was also confirmed by XRPD and e-PDF analysis. A broad diffuse peak observed in the XRPD pattern showed a shift toward higher angles following the higher reaction times of the ASR-treated material. A comparison of the EELS spectra revealed varying spectral features in the peak edges with different reaction times due to the interaction evolution between oxygen and the silicon and OH ions. Solid-state nuclear magnetic resonance (NMR) was also used to elucidate the silica nanostructures.

Keywords: amorphous silica; powder diffraction; transmission electron microscopy; high-resolution; spectroscopy; electron diffraction; electron pair distribution function

1. Introduction

SiO_2 (silica) is a three-dimensional siloxane bridged bond structure that is used as an aggregate or in nanocrystalline form, which, in recent decades, has been widely employed to create high-performance or highly functional materials [1]. The use of silica as an aggregate in silica glass has been intensively investigated for its specific heat insulation, good optical transmission, and high chemical resistance properties. Amorphous silica (a-silica) has various industrial (i.e., glass formers) and scientific applications, such as in photovoltaic cells and in electronic devices with optical properties [2–4]. Determining the response of porous silica to densification is challenging, as some amorphous materials are known to display anomalous behavior under high pressure [5]. Natural silica is used as an aggregate in composite materials such as concrete [6–8]. In fact, the degradation of concrete depends on the crystalline quality of the aggregate, where the reactivity of silica is dependent upon the chemical process that occurs between the amorphous or poorly crystallized silica present in mineral aggregates, referred to as an alkali silica reaction (ASR) [9–11].

Likewise, nanosilica is used in pottery clay materials as a strengthening additive, in electronic compounds as an insulator, and in the glass industry. It is also used to improve the creep resistance of thermoplastic polymers [12–15]. In contrast, the use of nanopowders presents a major health hazard, either during the manufacturing process or following the wear of a material, which may cause the release of nanoparticles into the environment. To solve this problem, it is of interest to manufacture nanostructured silica (<100 nm) in the form of arranged clusters. The advantage of this process is the low dimensionality properties of silica, while avoiding health risks during the manufacturing process following the wear of the material and its subsequent release into the environment. In addition to its technological relevance, the availability of amorphous nanostructured silica materials will enable the study of a vast range of interfacial phenomena and confined species.

Many previous studies have shown that the reactivity of silica compounds is due to the amorphous and strongly disordered part of silica. Different works have been carried out in order to characterize the nano- and micro-structures of concrete so as to improve its durability [16–18]. Micro-X-ray absorption near-edge structure (XANES) and micro-fluorescence experiments have been carried out to investigate the local structural evolutions of a heterogeneous and natural silica submitted to the ASR process [8]. Using micro-beam sources, micro-zones with different properties have been studied. Elemental maps obtained by environmental scanning transmission electron microscopy (ESEM) and micro-X-ray fluorescence (micro-XRF) demonstrate the accurate diffusion of potassium inside grains. Using Si K-edge XANES spectra has enabled the elucidation of the structural evolution induced by the alkali–silica reaction in silica from the outside to the inside of particles, showing no significant changes in the K cations [8,18–20]. However, some questions remain unanswered, for instance, regarding the contribution of different structural forms (i.e., amorphous and disordered) of silica in the ASR process.

Previous ASR studies on natural flint [8,18–21] have shown that the reaction begins with breaking the Si–O–Si bonds of the siloxane bridge and the formation of amorphous and nanocrystalline phases. However, the structural heterogeneity of flint, i.e., the presence of microcrystalline, nanocrystalline, and amorphous domains, complicates the study of degradation mechanisms and the reaction kinetics. All previously published studies have proposed incomplete structural models [16–21].

To explore and predict the properties and interfacial behavior of silicon, silica, and its hydrolysis, Van Din et al. [22,23] developed the "ReaxFF" reactive force field computational tool. In general, ReaxFF describes the breakdown and formation of bonds due to calculations of bonding states using interatomic distances [24]. The developed force field is empirical and bond order-dependent and requires fewer computer resources in comparison to methods based on quantum mechanics. The parameters of this force field were recently further developed to describe correctly the O migration mechanism in an Si network [25] at different temperatures (i.e., 880–2400 K).

The structural study of nanocrystalline amorphous materials is typically performed by neutron or X-ray diffraction [26]. However, the scattering cross-section of electrons is relatively large compared to neutrons or X-rays, making it easier to study nanovolumes using transmission electron microscope (TEM). The use of the electron pair distribution function (e-PDF) in TEM is ideal for studying amorphous materials where the acquisition time of e-PDF spectra is much shorter (milliseconds instead of hours) compared to neutron or X-ray diffraction. In contrast, the evolution of structures using a small quantity of a sample is more easily studied using electron diffraction and e-PDF analysis [27,28]. Currently, e-PDF is used only to extract short–long-range order information from nanoparticles, amorphous thin films, and amorphous organic materials. At present, there is not any work in the scientific bibliography related to extracting structural information on studying chemical reactions using e-PDF, which is an ideal method for studying structural changes in nanovolumes.

In this work, for the first time, we used a combination of high-resolution transmission electron microscopy (HRTEM), X-ray powder diffraction (XRPD), electron energy loss spectroscopy (EELS), electron pair distribution function (e-PDF), and solid-state nuclear magnetic resonance (NMR) to study the structural evolution of silica nanostructures with different reaction times during the ASR process. By combining different techniques, we were able to consistently correlate the morphology changes of amorphous silica nanostructures (information from HRTEM) with relation to Si-O environmental changes (information from EELS and e-PDF); in addition, we also confirmed those structural changes in bulk silica using XRPD. Short-range ordering (SRO) using e-PDF was also observed in the material even after several hours of hydrothermal reaction. In parallel, formation of Si-OH on the surface was also confirmed by NMR. In the following sections, we will present and discuss in detail our results related to the combined use of previously mentioned techniques.

2. Materials and Methods

The starting material used in this study was a-silica from Alfa Aesar (www.Alfa.com) (Ward Hill, MA, USA). The purity of silica was confirmed to be 99.9% by X-ray fluorescence analysis. The a-silica was submitted to the ASR process as previously described [20,21]. Briefly, 1 g of a-silica (S1) was submitted to an accelerated ASR process at 80 °C for 6 hours (S2), 168 hours (S3), and 312 hours (S4), with a mixture of 0.5 g of portlandite $Ca(OH)_2$ and 10 mL of potash solution KOH at 0.79 mol/L. The sample was retained under the ASR process for different periods in order to track the evolution of the resulting structure. Calcium and potassium were then removed by a selective acid treatment [20,21].

The XRPD spectra were recorded for 2θ values between 5° and 60° with steps of 0.007° and a counting time of 10 s per step using a Bruker D8 ADVANCE (Bruker AXS, Karlsruhe, Germany) diffractometer operating at 40 kV and 40 mA with Cu radiation (λCu = 0.15418 nm).

HRTEM and EELS measurements were carried out on a JEOL 3000FEG transmission electron microscope operating at 300 kV, equipped with a Gatan Enfina EELS spectrometer at the Electron Microscopy Centre, Madrid, Spain. The EELS energy resolution was approximately 1.2 eV for all spectra, as measured by the full-width at half-maximum (FWHM) of the corresponding zero-loss peak. Both the background and the plural scattering had to be subtracted from the experimental spectra to isolate the white line intensities. All analyzed crystals were very thin nanoparticles (mean free path $\lambda \leq 1$), where the EELS spectra were based on the fine edge of single nanoparticles to reduce the influence of multiple scattering effects.

For the electron diffraction (ED) measurements, the silica particles were dispersed onto the surface of carbon foil (20–30 nm thickness) that covered the Cu TEM grid (Tedpella, Redding, CA, USA). The ED measurements were performed using the nanoprobe mode of a Philips CM20 TEM (Philips Electron Optics division, Amsterdam, Netherlands) with an LaB6 cathode operating at 200 keV. The ED patterns were recorded on imaging plates (Ditabis), which provide a linear response to an electron dose over 6 orders of magnitude. We used the TEM nanoprobe mode (where the diameters of the studied and illuminated areas coincide) so as to avoid stray radiation from areas outside of the selected area (SA). This is essential in the e-PDF analysis of amorphous materials, as the signal from the amorphous structure is very weak—practically of the same order of magnitude as background—and therefore, even a small variation in intensity could greatly affect the result of the measurements. It is well known that nanocrystalline silica and hydrated silicates are highly sensitive to radiation, which undergo amorphization under an electron beam in a matter of seconds. To minimize beam damage [29], the ED patterns were obtained at a liquid nitrogen temperature and the incident electron beam intensity was kept as low as possible. The electron dose rate during the measurement was kept at the level of 10 e-/Å² s, which is far below the critical dose for this material.

The standard procedure we used to obtain reproducible ED patterns was as follows: (a) Start with the de-magnetization of all lenses, (b) set a fixed current for the objective lens, (c) position the studied area to the focal plane of the objective lens (using Z-control), and (d) set fixed values for the condenser lenses that focus the diffraction pattern [30–32]. For camera length calibration, self-supporting nanocrystalline Ni foil was used. Using this standard procedure, deviations of the camera length could be kept below 0.5%, regardless of the different samples examined. During the ED experiments, the beam diameter was set to ca. 1.5 µm, and for optimal background subtraction after recording the ED patterns from the silica particles, an additional ED pattern was recorded from the empty carbon foil under identical beam conditions. Radial averaging of the ED patterns was performed using process diffraction and the e-PDFSuite software [33–36]. The ED intensities were integrated radially into one-dimensional (1D) profiles and the background intensity from the carbon foil was subtracted from that of the silica particles. The resulting 1D ED profile, which contained intensity scattered exclusively by the silica particles, yielded input data for the e-PDF analysis.

Solid-state ^{29}Si magic angle spinning MAS NMR experiments were carried out on the Bruker Advance 100 spectrometer (Bruker BioSpin, Billerica, MA, USA) operating at magnetic fields of 2.34 T. The ^{29}Si MAS experiments, operated at 19.89 MHz, and the samples were spun in a cylindrical 7 mm ZrO$_2$ rotor at a spinning frequency of 4 kHz. ^{29}Si chemical shifts were determined relative to the tetramethylsilane as an external reference. The spectra were recorded with a pulse angle of π/5 and a recycle delay of 80 s, which was verified to enable relaxation. For each sample, a total of 256 scans were carried out.

3. Results and Discussion

Structural silica has SiO$_4$ tetrahedral units connected at the corners by bridging O. The ideal silica two-dimensional (2D) structures have linear Si–O–Si bonds, as illustrated in Figure 1, where the torsional energy required for one bond rotation depends on the bond angles that vary between 145° and 150° [37,38]. In the literature, the nature of Si–O–Si angles is extensively debated within the glass community, and an overview of the literature regarding measured and simulated Si–O–Si angles can be found in Reference [39]. Due to the large variety of Si–O–Si angles that join two neighboring building units, a-silica structures lack long-range order.

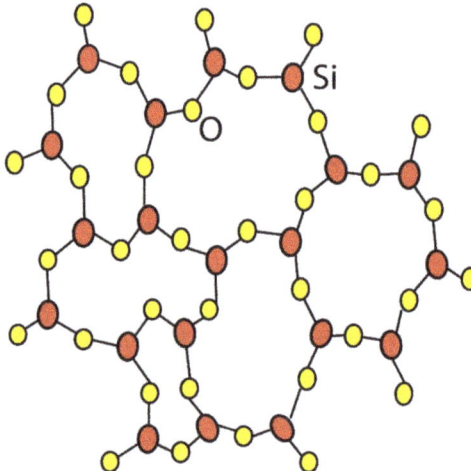

Figure 1. The ball-and-stick diagram of the structure of amorphous silica (adapted from Keen and Dove [33]).

Thereby, the local disorder and the orientation of SiO$_4$ tetrahedra could allow Si–O–Si angles to adjust to values that are more chemically stable. On a short-length scale, different phases may experience similar dynamic disorder regarding atom positions. In a recent study, we confirmed the hypothesis of the formation of silica clusters of Si–O with different structural states following the ASR process [40]. Depolymerization of the silica network creates Q4 (Si(OSi)$_4$) species, Q3 (Si(OSi)$_3$)(OH)) species that consist of a tetrahedral silicate sheet structure, Q2 species (SiO_4 tetrahedra in the middle of silicate chains), and Q1 species (tetrahedra at the end of silicates chain), as confirmed in a previous study [41]. Therefore, the degradation results of our material are not based on a simple formation of amorphous phases, but rather on a formation of nanodomains of heterogeneous sizes with different structures. In contrast, no high spatial resolution information exists about the electronic and chemical environment around silicon and oxygen within these structures.

The high-resolution electron microscopy (HREM) analysis of the samples was performed under the same magnification and the same defocus to better understand the microstructures of the amorphous state. Figure 2 shows the obtained micrographs studied for all compounds (i.e., S1 (a), S2 (b), S3 (c), and S4 (d)). We observed that two compounds (i.e., S1 and S2) were very stable under the beam, while the other two (i.e., S3 and S4) were slightly unstable after ~1 min of TEM. To avoid any artefact in the HRTEM images due to beam damage, data from each sample were collected under the same electron dose condition, far below the critical dose of the material. The images corresponding to samples S1 and S2 showed an amorphous microstructure, while those of S3 and S4 showed small changes compared to S1 and S2, since the presence of Si–Si ordering in a SRO with different arrangements of the Si–Si domains was observed (see enlarged images inserted in Figure 2c,d).

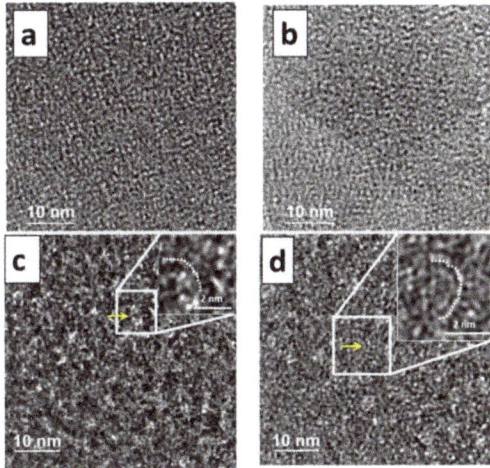

Figure 2. High-resolution electron microscopy images of the a-silica compounds S1 (**a**), S2 (**b**), S3 (**c**), and S4 (**d**). The area inside the white rectangle has been enlarged to show the ring structure. Enlarged images are inserted for S3 (**c**) and S4 (**d**), in which the nanopore ring formations are marked with white dots. Yellow arrows within the white square area indicate a nanopore ring formation.

All of the studied materials were found to be amorphous (Figure 2), which was confirmed also using XRPD (Figure 3) and subsequent e-PDF analysis (Figure 8). The HRTEM micrograph in Figure 2a shows the morphology of the starting amorphous silica. The evolution of the silica ring formation network can be observed in S3 and S4, and in Figure 2b–d, the morphology of the amorphous silica can be observed after the reaction. An increase in the size of the nanopore ring formation marked with white dots in the inserts of Figure 2c, d in the silica network can be more clearly observed for S4 than for S3. In addition, the nanopore distribution is highly heterogeneous in S3 and S4, leading to the formation of short-range ordering only. As an important remark, it is interesting to note that the number of Si atoms that form part of the Si tetrahedral ring is lower in S3 than in S4, as indicated by the white dots in Figure 2c,d (magnified insert).

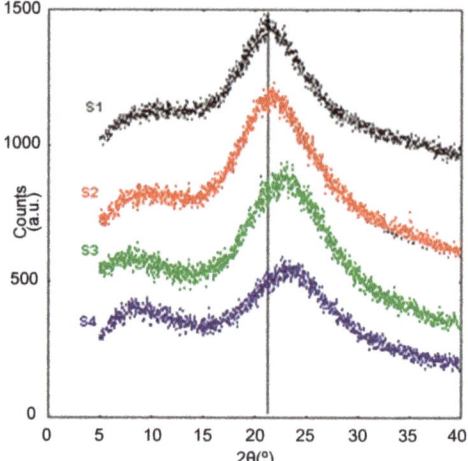

Figure 3. The X-ray powder diffraction (XRPD) spectra of a-silica (S1–S4) using various alkali silica reaction (ASR) times, where the x-axis is 2θ and the y-axis is the intensity in arbitrary units. The vertical line represents the peak maxima of the starting material.

In Figure 3, the XRPD pattern of amorphous silica and subsequent reaction materials shows typical diffuse peaks, confirming the presence of an amorphous structure and/or very small crystal sizes [42]. For the starting material (S1), a broad diffuse peak maxima is located at 2θ = 21°, which is a well-known feature for amorphous silica material. A subsequent shift of the maximum is observed toward the higher angles with increasing ASR reaction times.

As observed during the HRTEM image acquisition, the particles are of nanometer scale, producing broad Bragg peaks in the XRPD data for the studied materials. With the ASR process, it is possible that the Q4 species convert to Q3, Q2, and Q1 species. As the reaction time increases, the Q1 and Q2 species may form, creating increased diversity in the silica structure bond distance distribution, with broader diffuse peaks in the XRPD data.

With higher reaction times, the average bond distance in the structure becomes shorter, possibly due to more O–H species connected to Si, which translates into a shift of the X-ray diffraction peaks toward higher angles, as observed in the XRPD data for S4. The change in structure, as observed in the XRPD patterns, is consistent with the HRTEM images.

To shed more light on the structural changes, EELS analysis was performed to observe variations in the detailed features around the Si–L and O–K edges. The obtained EELS spectra were compared with EELS reference bibliography data [43] to check for possible energy shifts or any variation in the shape of the edges.

It is believed that the incorporation of OH⁻ groups into the amorphous silica system generates a mixed Si electronic configuration. Therefore, for atoms such as Si, the $L_{2,3}$ ratio might be expected to vary following electronic configuration changes. To establish the formal electronic configuration and/or oxidation state of Si in each composition/reaction time, EELS analysis was performed by comparing the evolution of the Si $L_{2,3}$ edges for the whole studied range/reaction times (S1–S4), as shown in Figure 4.

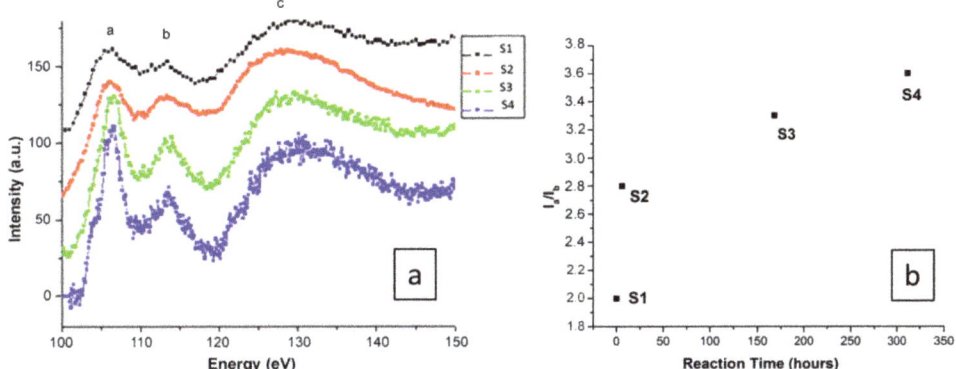

Figure 4. (a) Electron energy loss spectra of the Si $L_{2,3}$ edges obtained for all compounds/reaction times. Data were normalized to the L_3 maximum intensity. (b) Evolution of the intensity ratio of the peaks a to b deduced from the Si $L_{2,3}$ electron energy loss spectroscopy (EELS) spectra.

The EELS spectra corresponding to S1–S4 are shown in Figure 4a. The Si $L_{2,3}$ edges for the studied compounds were normalized to the L3 maximum intensity. The energy loss near-edge spectra (ELNES) show a first defined peak around 107 eV (marked a), followed by a broad peak at 114 eV (marked b), and then a very broad peak with a maximum at 130 eV (marked c). The results show that although Si–$L_{2,3}$ core loss edges have an appropriate energy loss range (approximately 108 eV) [43], and the peaks are located at similar positions in the four spectra, the relative peak intensities change dramatically from S1 to S4. Although Si peaks can be observed for all samples, they change from broad (S1) to sharp (S4) peak positions as the reaction time increases. Such a difference in the Si–$L_{2,3}$ peaks is probably due to Si tetrahedral distortion, since in the SiO_4 tetrahedra, the four distances between the silicon and the oxygens are slightly different. The higher reaction times indicate that the OH^- ions were incorporated into the silica amorphous structure, which may be associated with the increase in the white line Si–$L_{2,3}$ intensity, and generally, the spectra become better defined with sharper EELS peaks. The appearance of a sharper EELS peak as the reaction time increases is probably due to the incorporation of OH^- around Si atoms, leading to more regular (less distorted) Si tetrahedra [43]. A small shoulder observed in S4 indicates that some of the Si tetrahedra remain distorted in the measured sample, since the perfect maxima are only observed if all of the Si tetrahedra are undistorted [44]. The EELS spectra show strong modifications of the silica tetrahedral network at different reaction times. The ratio between peaks a to b deduced from the Si–$L_{2,3}$ EELS are shown in Figure 4b, highlighting that instead of a linear increase, hydration happens very fast with a huge distortion in the Si tetrahedra in the first instance, and then a tendency towards plateaus.

We also studied the evolution of the O–K edges of the EELS spectra with different reaction sample species (S1–S4) to provide information on the coordination structure of local oxygen atoms, such as the configuration and the type of neighboring species. The O–K edges in the spectrum were caused by the transition from the O 1s state to the O 2p final state in the conduction band, hybridized with the valence orbitals of neighboring atoms. The features of the O–K edges of all compounds were found to be quite similar (Figure 5).

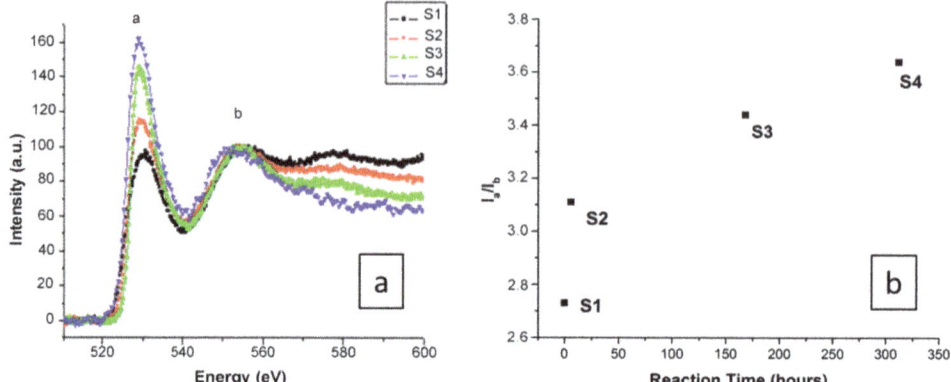

Figure 5. (**a**) Electron energy loss spectra taken of the O–K edges of all compounds. All of the spectra were calibrated at the O–K pre-edge peak position. (**b**) Evolution of the intensity ratio of peaks a to b deduced from the O–K electron energy loss spectroscopy (EELS) spectra.

Generally, the sharpness of peak "a" of the O–K edges arose from O–O scattering and increased in intensity (Figure 5a) as the number of O second-nearest neighbors around the excited O atom increased [39]. The results show that although the O–K edges have an appropriate energy loss range (approximately 532 eV) [45,46] with peaks located at similar positions in the four spectra, the relative intensity of the peaks changes from S1 to S4 with different reaction times, where the observed increase in the O–K peak intensity is related to the presence of additional oxygen atoms around the excited oxygen atoms. The EELS spectra therefore show strong modifications regarding the oxygen environment at different reaction scales from samples S1 to S4. The ratio between peaks a to b deduced from the O–K EELS spectra (Figure 5b) shows a tendency towards a plateau instead of a linear increase with reaction time, which can also be observed in the EELS spectra of Si (Figure 4b).

Such an increase could be correlated with an accelerated hydrolysis reaction of the OH$^-$ groups with silica from S1 to S4, in agreement with the infrared (IR) spectroscopy results published by Hamoudi et al. [47]. Figure 6 provides a schematic of the amorphous silica–OH$^-$ interaction, where the reaction mechanism starts at the surface and, via a stepwise mechanism, hydrates the silica, in a similar way as proposed previously by Dove et al. [48] in the case of quartz–water interactions.

To complete the results obtained by HRTEM, EELS, and XRPD, a ^{29}Si MAS-NMR experiment was performed in order to obtain information about short-range order changes and about the surface of silica nanostructures. NMR experiments were performed only for two samples with very different behaviors in order to check the creation of Q3 species. Figure 7 shows the ^{29}Si NMR-MAS corresponding to the starting form of silica (S1) and silica after a reaction time of 312 hours (S4). In the starting form of silica (S1), the major species are Q4, corresponding to amorphous silica with SiO$_4$ tetrahedral units, each one connected to four tetrahedra via oxygen. The spectrum for S1 presents a line, centered around −110 ppm, that is attributed to Q$_4$ species, which has also been confirmed by previous studies [49–51].

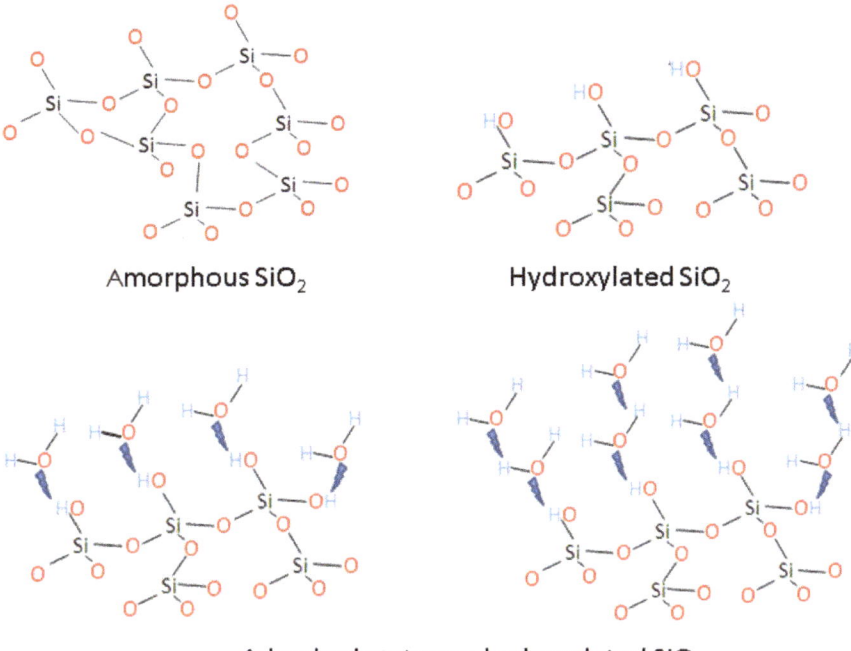

Figure 6. Schematic diagram illustrating stepwise OH⁻ interactions with a-silica. Hydrogen bonding between absorbed water molecules is not shown in the figure, as it is beyond the scope of our study.

For the final reaction material, S4, in addition to Q4 species, a second line centered around −101 ppm can be observed, which can be attributed to the Q3 species that correspond to silanol groups Si–OH. The formation and changes of the Si–OH species and Si–O–Si bonds were also confirmed previously in a similar material using Fourier transform infrared spectroscopy [47]

From the collected ED data (samples S1–S4), the electron pair distribution function (G(r)), which provides a measure of the probability of finding two atoms separated by distance r (Figure 8), was calculated using the e-PDFSuite and the Process Diffraction software [34–36,52] developed to analyze the ED patterns of amorphous and nanocrystalline materials. During the calculation of the 1D distribution from the 2D electron diffraction patterns of silica materials, the contribution of amorphous carbon support was subtracted where the distortions in the 2D diffraction patterns were corrected. The beam stopper area and the dead pixels in the detector were masked and eliminated from the ED pattern during e-PDF calculations. The detailed procedure of the G(r) calculation from the 1D distribution data using the e-PDFSuite and Process Diffraction software program has been described elsewhere [34–36].

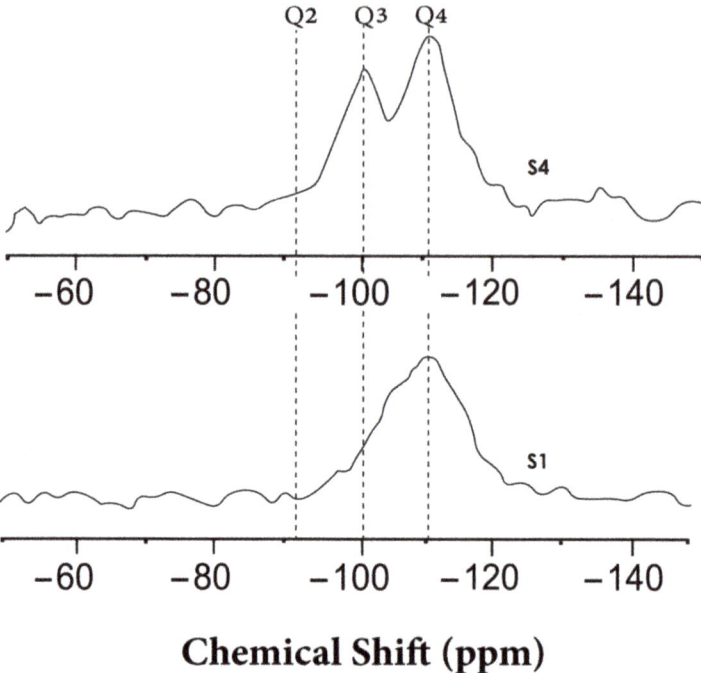

Figure 7. Local ordering of the starting form of a-silica S1 and after reaction time of 312 hours (S4): ^{29}Si MAS NMR spectra, where the x-axis shows the chemical shift in parts per million (ppm) and the y-axis corresponds to the intensity scale of each spectral line in arbitrary units.

Figure 8a shows that, following the e-PDF analysis, no peak (corresponding to interatomic distances) can be found beyond 8 Å in any of the hydrothermal products (S1–S4). This fact confirms that only short-range ordering is present in the material, even after several hours of hydrothermal reaction. It is interesting to note that the minor ripples observed in the e-PDF results beyond 8 Å are due to the limited Q range resolution and the attenuation of the signal amplitude (obtained Q = 14 Å^{-1}, where Q = 2Π/d), and they do not contain any structural information. In the 1.3–3.5 Å region for S1–S4, small changes can be observed in the e-PDF peak positions (corresponding to the interatomic Si–Si, Si–O, and O–O distances). The small peak at 2.0 Å is probably an artifact, due to limited Q resolution or due to additional Si–O connectivity in the amorphous state. All such peaks/interatomic distances in the 1.3–3.5 Å region match well with the distances that exist within the silica crystalline structure [53], as shown in Figure 8b. As a consequence of the precision of the peak localization in the e-PDF analysis decreasing with increasing atomic distances due to the enhanced peak broadening, the relatively sharp first, second, and third e-PDF peaks corresponding to the nearest interatomic distances can be reliably considered for our comparison (Figure 8b). According to the results, small changes (i.e., shortening of bond distances) can be observed for species with higher reaction times (more obvious for S4 in comparison to S1), which confirm our previous observation from the XRPD patterns and EELS spectra of the Si and O edges.

The interatomic distances extracted from the e-PDF calculation are accurate, as the ED pattern calibration and the determination of the center of the ED pattern may cause an error smaller than ± 0.005 Å within the range of 1–5 Å [54–56]. To establish high accuracy of the observed peak positions, during our study, a proper calibration of the TEM camera length was carried out, and all the experiments were performed under similar microscope conditions. Electron diffraction is subject to multiple scattering, and this multiple scattering is prominent for nanomaterials with higher particle sizes. However, it was demonstrated by Anstis et al. [57] that multiple scattering does not affect peak positions in $G(r)$ if $t/\alpha \leq 5$ (where t is the sample thickness and α is the elastic mean free path), although it may affect coordination number determination. On the other hand, as a-silica consists of only light elements, we did not consider correcting for multiple scattering during the $G(r)$ calculation.

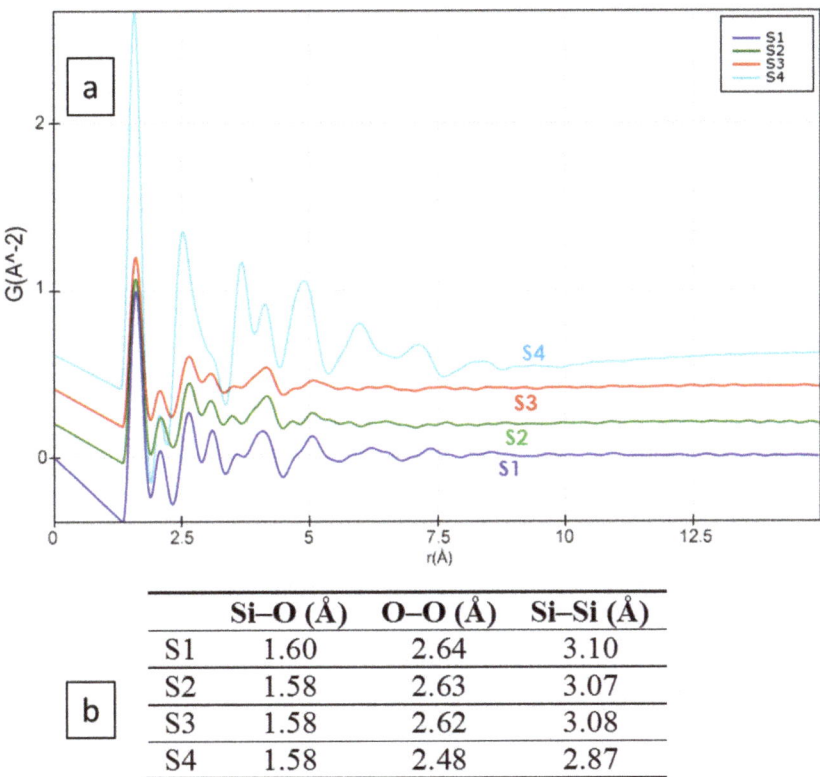

Figure 8. (a) The calculated electron pair distribution function (e-PDF) from S1–S4, where the x-axis represents the interatomic distances in Å and the y-axis represents the probability (in arbitrary units) of finding two atoms at distance r. (b) Calculated of the peak position from the e-PDF analysis.

4. Conclusions

The study of amorphous silica and its hydrothermal reactions at different time intervals by using various complementary techniques—namely, HRTEM, XRPD, NMR, and e-PDF—revealed the amorphous nature of the nanostructure of the resulting material. All of the studied samples were found to be amorphous, as observed by HRTEM, a fact that was also confirmed by XRPD and e-PDF analysis.

The combined results of these techniques shed light on the structural changes that occur at different reaction times. A shift of the diffuse peak maxima toward higher angles in the XRPD patterns indicates shortening of the average bond distance, a fact that was also confirmed by e-PDF analysis. The changes (i.e., shortening of bond distances) for species with higher reaction times observed via e-PDF analysis were also confirmed by the XRPD and HRTEM results.

The observed structural changes might be due to the Si–O tetrahedral contraction with increasing reaction times, which were also observed in the EELS spectra and could be explained by the depolymerization of the amorphous structure and the presence of some OH species, as also confirmed by NMR spectroscopy. The NMR results also revealed the presence of Si–OH species, indicating the existence of a three-dimensional network of silica. For the final reaction product (S4), the solid-state NMR analysis showed that, in addition to Q4 species, a second line centered around −101 ppm appeared, which can be attributed to the Q3 species that appeared in the silica nanostructures, thus confirming the results obtained using the EELS technique.

Again, our current results are in agreement with a recent study concerning the densification effects on porous silica using molecular dynamics, where the rearrangement of Si–O–Si is part of an important structural change effect during densification. In addition, densification has been shown to produce more rings in vitreous silica, which could be attributed to the effect of repolymerization [5].

Finally, this study also showed the importance of using many combined state-of-the-art techniques to study complex structural problems related to the reactivity of amorphous silica. RMC (Reverse Monte Carlo) modeling of the obtained e-PDF data are in progress in order to provide additional insight about the amorphous silica structure.

Author Contributions: Experimental procedure and the data processing, L.K., K.B., P.P.D., V.K.K., and J.L.L.; designing the experiment, L.K., K.B., P.P.D., S.N., V.K.K., and J.L.L. All authors participated in preparing the manuscript. All authors have read and agreed to the published version of the manuscript.

Funding: Part of this research was funded by Spanish Ministry of Economy and Competitiveness, grant number MAT2016- 78362-C4-4-R.

Conflicts of Interest: The authors declare no conflict of interest.

References

1. Zhang, W.; Blackburn, R.S.; Dehghani-Sanij, A.A. Effect of silica concentration on electrical conductivity of epoxy resin–carbon black–silica nanocomposites. *Scr. Mater.* **2007**, *56*, 581–584. [CrossRef]
2. Ojha, N.; Trautvetter, T.; Norrbo, I.; Kalide, A.; Lastusaari, M.; Mueller, R.; Petit, L. Sintered silica bodies with persistent luminescence. *Scr. Mater.* **2019**, *166*, 15–18. [CrossRef]
3. Fanderlik, I. *Silica Glass and Its Application*, 1st ed.; Elsevier: Amsterdam, The Netherlands, 1991; ISBN 9781483291680.
4. Mauro, J.C.; Zanotto, E.D. Two Centuries of Glass Research: Historical Trends, Current Status, and Grand Challenges for the Future. *Int. J. Appl. Glas. Sci.* **2014**, *5*, 313–327. [CrossRef]
5. Tian, Y.; Du, J.; Hu, D.; Zheng, W.; Han, W. Densification effects on porous silica: A molecular dynamics study. *Scr. Mater.* **2018**, *149*, 58–61. [CrossRef]
6. Meral, C.; Benmore, C.J.; Monteiro, P.J. The study of disorder and nanocrystallinity in C–S–H, supplementary cementitious materials and geopolymers using pair distribution function analysis. *Cem. Concr. Res.* **2011**, *41*, 696–710. [CrossRef]
7. Dharmawardhana, C.C.; Misra, A.; Aryal, S.; Rulis, P.; Ching, W.Y. Role of interatomic bonding in the mechanical anisotropy and interlayer cohesion of CSH crystals. *Cem. Concr. Res.* **2013**, *52*, 123–130. [CrossRef]
8. Khouchaf, L.; Boinski, F.; Tuilier, M.-H.; Flank, A.M. Characterization of heterogeneous SiO2 materials by scanning electron microscope and micro fluorescence XAS techniques. *Nucl. Instrum. Methods Phys. Res. Sect. B (NIMB)* **2006**, *252*, 333–338. [CrossRef]
9. Chatterji, S. Chemistry of Alkali-Silica Reaction and Testing of Aggregates. *Cem. Concr. Compos.* **2005**, *27*, 788–795. [CrossRef]
10. Dent Glasser, L.S.; Kataoka, N. The Chemistry of 'Alkali-Aggregate' Reaction. *Cem. Concr. Res.* **1981**, *11*, 1–9. [CrossRef]
11. Broekmans, M.A.T.M. Structural properties of quartz and their potential role for ASR. *Mater. Charact.* **2004**, *53*, 129–140. [CrossRef]
12. Mačkovića, M.; Niekiela, F.; Wondraczek, L.; Bitzek, E.; Spiecke, E. Scripta Materialia. In situ mechanical quenching of nanoscale silica spheres in the transmission electron microscope. *Scr. Mater.* **2016**, *121*, 70–74. [CrossRef]
13. Wondraczek, L.; Mauro, J.C.; Eckert, J.; Kühn, U.; Horbach, J.; Deubener, J.; Rouxel, T. Towards Ultrastrong Glasses. *Adv. Mater.* **2011**, *23*, 4578–4586. [CrossRef] [PubMed]

14. Kermouche, G.; Barthel, E.; Vandembroucq, D.; Dubujet, P. Mechanical modelling of indentation-induced densification in amorphous silica. *Acta Mater.* **2008**, *56*, 3222–3228. [CrossRef]
15. Zhou, T.H.; Ruan, W.H.; Yang, J.L.; Rong, M.Z.; Zhang, M.Q.; Zhang, Z. A novel route for improving creep resistance of polymers using nanoparticles. *Compos. Sci. Technol.* **2007**, *67*, 2297–2302. [CrossRef]
16. Khouchaf, L.; Hamoudi, A.; Cordier, P. Evidence of depolymerisation of amorphous silica at medium and short range order: XANES, NMR and CP-SEM contributions. *J. Hazard. Mater.* **2009**, *168*, 1188–1191. [CrossRef] [PubMed]
17. Balas, F.; Rodríguez-Delgado, M.; Otero-Arean, C.; Conde, F.; Matesanz, E.; Esquivias, L.; Ramírez-Castellanos, J.; Gonzalez-Calbete, J.; Vallet-Regí, M. Structural characterization of nanosized silica spheres. *Solid State Sci.* **2007**, *9*, 351–356. [CrossRef]
18. Khouchaf, L.; Boinski, F. Environmental scanning electron microscope study of SiO_2 heterogeneous material with helium and water vapor. *Vacuum* **2007**, *81*, 599–603. [CrossRef]
19. Khouchaf, L.; Verstraete, J.; Prado, R.J.; Tuilier, M.H. XANES, EXAFS and RMN contributions to follow the structural evolution induced by alkali-silica reaction in SiO_2 aggregate. *Phys. Scr.* **2005**, *115*, 552. [CrossRef]
20. Khouchaf, L.; Verstraete, J. Multi-technique and multi-scale approach applied to study the structural behavior of heterogeneous materials: Natural SiO_2 case. *J. Mater. Sci.* **2007**, *42*, 2455–2462. [CrossRef]
21. Verstraete, J.; Khouchaf, L.; Tuilier, M.H. Contributions of the Environmental Scanning Electron Microscope and X-ray diffraction in investigating the structural evolution of a SiO_2 Aggregate attacked by Alkali-Silica Reaction. *J. Mater. Sci.* **2004**, *39*, 6221–6226. [CrossRef]
22. Goddard, W.A.; Van Duin, A.C.T.; Strachan, A.; Stewman, S.; Zhang, Q.; Xu, X. ReaxFFSiO reactive force field for silicon and silicon oxide systems. *J. Phys. Chem. A* **2003**, *107*, 3803–3811. [CrossRef]
23. Senftle, T.; Hong, S.; Islam, M.; Kylasa, S.B.; Zheng, Y.; Shin, Y.K.; Junkermeier, C.E.; Engel-Herbert, R.; Janik, M.J.; Aktulga, H.M.; et al. The ReaxFF reactive force-field: Development, applications and future directions. *NPJ Comput. Mater.* **2016**, *2*, 2057–3960. [CrossRef]
24. Fogarty, J.C.; Aktulga, H.M.; Grama, A.Y.; van Duin, A.C.T.; Pandit, S.A. A reactive molecular dynamics simulation of the silicawater interface. *J. Chem. Phys.* **2010**, *132*, 174704. [CrossRef] [PubMed]
25. Nayir, N.; Van Duin, A.C.T.; Erkoc, S. Development of the ReaxFF Reactive Force Field for Inherent Point Defects in the Si/Silica System. *J. Phys. Chem. A* **2019**, *123*, 4303–4313. [CrossRef]
26. Egami, T.; Billinge, S.J.L. *Underneath the Bragg Peaks: Structural Analysis of Complex Materials*; Pergamon Press: Oxford, UK; Elsevier: Amsterdam, The Netherlands, 2003.
27. Abeykoon, M.; Malliakas, C.D.; Juhás, P.; Božin, E.S.; Kanatzidis, M.G.; Billinge, S.J.L. Quantitative nanostructure characterization using atomic pair distribution functions obtained from laboratory electron microscopes. *Z. für Krist.* **2012**, *227*, 248–256. [CrossRef]
28. Cockayne, D.J.H.; McKenzie, D.R. Electron diffraction analysis of polycrystalline and amorphous thin films. *Acta Cryst. A* **1988**, *44*, 870–878. [CrossRef]
29. Egerton, R.F.; Li, P.; Malac, M. Radiation damage in the TEM and SEM. *Micron* **2004**, *35*, 399–409. [CrossRef]
30. Lábár, J.L. Electron Diffraction Based Analysis of Phase Fractions and Texture in Nanocrystalline Thin Films, Part I: Principles. *Microsc. Microanal.* **2008**, *14*, 287–295. [CrossRef]
31. Lábár, J.L. Electron Diffraction Based Analysis of Phase Fractions and Texture in Nanocrystalline Thin Films, Part II: Implementation. *Microsc. Microanal.* **2009**, *15*, 20–29. [CrossRef]
32. Abeykoon, M.; Hu, H.; Wu, L.; Zhu, Y.; Billinge, S.J.L. Calibration and data collection protocols for reliable lattice parameter values in electron pair distribution function studies. *J. Appl. Cryst.* **2015**, *48*, 244–251. [CrossRef]
33. e-PDFSuite–Software for Analysis of Amorphous and Nano Crystalline Materials. NanoMEGAS SPRL: Belgium, 2009. Available online: http://www.nanomegas.com (accessed on 30 September 2020).
34. Lábár, J.L.; Das, P.P. Pattern Center and Distortion Determined from Faint, Diffuse Electron Diffraction Rings from Amorphous Materials. *Microsc. Microanal.* **2007**, *23*, 647–660. [CrossRef] [PubMed]
35. Hoque, M.M.; Vergara, S.; Das, P.P.; Ugarte, D.; Santiago, U.; Kumara, C.; Whetten, R.L.; Dass, A.; Ponce, A. Structural Analysis of Ligand-Protected Smaller Metallic Nanocrystals by Atomic Pair Distribution Function under Precession Electron Diffraction. *J. Phys. Chem. C* **2019**, *123*, 19894–19902. [CrossRef]
36. Lábár, J.L. Electron Diffraction Based Analysis of Phase Fractions and Texture in Nanocrystalline Thin Films, Part III: Application Examples. *Microsc. Microanal.* **2012**, *18*, 406–420. [CrossRef] [PubMed]

37. Keen, D.A.; Dove, M.T. Local structures of amorphous and crystalline phases of silica, SiO2, by neutron total scattering. *J. Phys. Condens. Matter.* **1999**, *11*, 9263–9273. [CrossRef]
38. Awazu, K.; Kawazoe, H. Strained Si—O—Si Bonds in Amorphous SiO2 Materials: A Family Member of Active Centers in Radio, Photo, and Chemical Responses. *Appl. Phys. Rev.* **2003**, *94*, 6243–6262. [CrossRef]
39. Yuan, X.; Cormack, A. Si–O–Si bond angle and torsion angle distribution in vitreous silica and sodium silicate glasses. *J. Non-Cryst. Solids* **2003**, *319*, 31–43. [CrossRef]
40. Boinski, F.; Khouchaf, L.; Tuilier, M.-H. Study of the mechanisms involved in reactive silica. *Mater. Chem. Phys.* **2010**, *122*, 311–315. [CrossRef]
41. Bednarik, V.; Melar, J.; Vondruska, M.; Slavik, R. Influence of Silicate Depolymerisation on the Polycondensation Reaction with Hydroxoaluminate in Alkaline Aqueous Solution. *J. Inorg. Organomet. Polym. Mater.* **2011**, *21*, 9–14. [CrossRef]
42. Zhang, G.; Xu, Y.; Xu, D.; Wang, D.; Xue, Y.; Su, W. Pressure-induced crystallization of amorphous SiO2 with silicon–hydroxy group and the quick synthesis of coesite under lower temperature. *High Press. Res.* **2008**, *28*, 641–650. [CrossRef]
43. Ahn, C.C.; Krivanek, O.L.; Disko, M.M. *EELS Atlas: A Reference Guide of Electron Energy Loss Spectra Covering All Stable Elements*; ASU HREM Facility and GATAN, Center for Solid State Science, Arizona State University: Tempe, AZ, USA, 1983.
44. Jiang, N.; Silcox, J. Observations of reaction zones at chromium oxide glass interfaces. *J. Appl. Phys.* **2000**, *87*, 3768–3775. [CrossRef]
45. Wallis, D.; Gaskell, P.H.; Brydson, R. Oxygen K near-edge spectra of amorphous silicon suboxides. *J. Microsc.* **1993**, *180*, 307–312. [CrossRef]
46. Muller, D.; Sorsch, T.; Moccio, S.; Baumann, F.H.; Evans-Luteeerodt, K.; Timp, G. The electronic structure at the atomic scale of ultrathin gate oxides. *Nature* **1999**, *399*, 758–761. [CrossRef]
47. Hamoudi, A.; Khouchaf, L.; Depecker, C.; Revel, B.; Montagne, L.; Cordier, P. Microstructural evolution of amorphous silica following alkali–silica reaction. *J. Non-Cryst. Solids* **2008**, *354*, 5074–5078. [CrossRef]
48. Dove, P.D.; Rimstidt, J.D. Silica-water interactions. *Rev. Mineral. Geochem.* **1994**, *29*, 259–308.
49. Kim, H.N.; Lee, S.K. Atomic structure and dehydration mechanism of amorphous silica: Insights from 29Si and 1H solid-state MAS NMR study of SiO2 nanoparticles. *Geochim. Cosmochim. Acta* **2013**, *120*, 39–64. [CrossRef]
50. Brinker, C.J.; Kirkpatrick, R.J.; Tallant, D.R.; Bunker, B.C.; Montez, B.J. NMR confirmation of strained "defects" in amorphous silica. *J. Non-Crys. Solids* **1988**, *99*, 418–428. [CrossRef]
51. Cody, G.D.; Mysen, B.O.; Lee, S.K. Structure vs. composition: A solid-state 1H and 29Si NMR study of quenched glasses along the Na2O-SiO2-H2O join. *Geochim. Cosmochim. Acta* **2005**, *69*, 2373–2384. [CrossRef]
52. Jian-Sheng, J.; Fang-Hua, L. Fitting the atomic scattering factors for electrons to an analytical formula. *Acta Phys. Sin.* **1984**, *33*, 845–849. [CrossRef]
53. Kis, V.K.; Dódony, I.; Lábár, J.L. Amorphous and partly ordered structures in SiO2 rich volcanic glasses. An ED study. *Eur. J. Mineral.* **2006**, *18*, 745–752. [CrossRef]
54. Billinge, S.J.L.; Farrow, C.L. Towards a robust ad hoc data correction approach that yields reliable atomic pair distribution functions from powder diffraction data. *J. Phys. Condens. Matter.* **2013**, *25*, 454202. [CrossRef]
55. Mu, X.; Wang, D.; Feng, T.; Kübel, C. Radial distribution function imaging by STEM diffraction: Phase mapping and analysis of heterogeneous nanostructured glasses. *Ultramicroscopy* **2016**, *168*, 1–6. [CrossRef] [PubMed]
56. Mu, X. TEM Study of the Structural Evolution of Ionic Solids from Amorphous to Polycrystalline Phases in the Case of Alkaline Earth Difluoride Systems Experimental Exploration Of Energy Landscape. Ph.D. Thesis, Karlsruhe Institute of Technology, Karlsruhe, Germany, 2013.
57. Anstis, G.R.; Liu, Z.; Lake, M. Investigation of amorphous materials by electron diffraction The effects of multiple scattering. *Ultramicroscopy* **1988**, *26*, 65–69. [CrossRef]

© 2020 by the authors. Licensee MDPI, Basel, Switzerland. This article is an open access article distributed under the terms and conditions of the Creative Commons Attribution (CC BY) license (http://creativecommons.org/licenses/by/4.0/).

Article

Correlative Light and Transmission Electron Microscopy Showed Details of Mitophagy by Mitochondria Quality Control in Propionic Acid Treated SH-SY5Y Cell

Minkyo Jung [1,†], Hyosun Choi [1,2,†], Jaekwang Kim [3] and Ji Young Mun [1,*]

1. Neural Circuit Research Group, Korea Brain Research Institute, Daegu 41062, Korea; j0312@kbri.re.kr (M.J.); hyokchoi0123@gmail.com (H.C.)
2. BK21 Plus Program, Department of Senior Healthcare, Graduate School, Eulji University, Daejeon 34824, Korea
3. Dementia Research Group, Korea Brain Research Institute, Daegu 41062, Korea; kim_jaekwang@kbri.re.kr
* Correspondence: jymun@kbri.re.kr
† These authors contributed equally to this work.

Received: 25 August 2020; Accepted: 27 September 2020; Published: 29 September 2020

Abstract: Propionic acid is a metabolite of the microbiome and can be transported to the brain. Previous data show that propionic acid changes mitochondrial biogenesis in SH-SY5Y cells and induces abnormal autophagy in primary hippocampal neurons. Maintaining mitochondrial function is key to homeostasis in neuronal cells, and mitophagy is the selective autophagy involved in regulating mitochondrial quality. Monitoring mitophagy though light microscopy or conventional transmission electron microscopy separately is insufficient because phases of mitophagy, including autophagosome and autolysosome in nano-resolution, are critical for studies of function. Therefore, we used correlative light and electron microscopy to investigate mitochondrial quality in SH-SY5Y cells after propionic acid treatment to use the advantages of both techniques. We showed, with this approach, that propionic acid induces mitophagy associated with mitochondrial quality.

Keywords: propionic acid; autophagy; mitophagy; correlative light and electron microscopy (CLEM)

---

1. Introduction

Short-chain fatty acids (SCFAs) such as acetic, propionic, and butyric acid are by-products of fermentation of dietary fiber by the gut microbiome [1]. Microbe-derived metabolites can cross the blood–brain barrier and affect the neurons. As the relationship between gut microbiome and the brain, gut–brain axis, has become interesting, SCFAs have attracted increasing attention. Propionic acid (PPA) is increased in stools from patients with autistic spectrum disorder, and prenatal exposure to PPA causes significant impairment of the social behavior of neonatal rat offspring [2]. Further, PPA administration to rodents alters expression of genes associated with neurotransmitters, neuronal cell adhesion molecules, inflammation, oxidative stress, lipid metabolism, and mitochondrial function [3–5]. Conversely, decreases in PPA are reported in patients with multiple sclerosis, an autoimmune and neurodegenerative disease [6]. One important cellular process negatively affected by PPA is mitochondrial function [7]. Rats exposed to PPA show mitochondrial dysfunction and an increase in free acyl-carnitine, a factor for the transport of long-chain fatty acids into mitochondria [8]. PPA and butyric acid also induce autophagy in human colon cancer cells that limits apoptosis, and inhibition of autophagy potentiates SCFA-induced apoptosis [9]. As our previous data indicated that PPA induces abnormal autophagy in PPA-treated hippocampal neuron [10], we investigated the relationship between mitochondrial defects and the regulation of mitochondrial quality though mitophagy.

Mitophagy is a selective degradative process responsible for removing damaged mitochondria to maintain cytoplasmic homeostasis [11]. Mitochondrial dysfunction is involved in various neurodegenerative or neurodevelopmental diseases [12]. Once mitophagy is initiated, a balance between autophagosome formation and autophagic degradation is necessary. Thus, accumulation of autophagosomes and disruption of the autophagic process in neurons is associated with disease [13]. Until now, conventional techniques for analysis of mitophagy have been based on immunofluorescence staining and immunoblotting of several specific mitochondrial proteins, qPCR for mitochondrial DNA copy number, and nano-resolution imaging using transmission electron microscopy (TEM). TEM is a direct imaging method for the early stage of mitophagy, which is the starting point of engulfing mitochondria and early autophagosome showing specific mitochondrial structures such as cristae [14,15]. However, the assessment of the late phases of mitophagy requires specific imaging techniques.

Recently, the engineering of two fluorescent proteins (mCherry-GFP-mito, and mt-Keima) has permitted monitoring of the status of mitophagy in live cells. These reporters change the fluorescence profile in response to pH changes. For example, the excitation wavelength for mt-Keima is 488 at neutral pH and 561 at acidic pH for late mitophagy observed in the lysosome [16,17]. mCherry-GFP-mito protein, fused to a mitochondrial targeting sequence of a mitochondrial protein, such as the outer mitochondrial membrane (OMM) protein FIS1 (comprising amino acids 101–152) [18] can be used to detect mitophagy. The mitochondrial network can be seen as a green fluorescence, and mitochondria delivered to lysosomes show as a red color after mitophagy. This is because mCherry fluorescence is stable, but GFP fluorescence is quenched in the acidic condition. However, these tools do not allow the monitoring of all phases of mitophagy. To analyze the entire dynamic phase of autophagy regarding mitophagy, TEM is employed to classify the specific type of autophagy including phagophore, autophagosome, and autolysosome in high resolution [19,20]. Therefore, we analyzed structural changes by TEM to study specific stages of autophagy that are more tightly linked with the mechanisms of mitophagy dysfunction. Thus, correlative light and electron microscopy (CLEM) is an effective method to analyze mitophagy or autophagic pathways [21]. Because the Keima protein is incompatible with fixation [22], we used GFP and mCherry conversion depending on the pH level of lysosomes to investigate details of various steps of mitophagy in nano-resolution though electron microscopy (EM). The CLEM technique of overlaying two images from fluorescence and EM makes the investigation of all phases of mitophagy possible. Evans et al. suggested that CLEM can open new avenues using light-up through (fluorescent) dyes in the dark by EM observation [23]. Thus, we applied CLEM to study mitophagy after PPA treatment.

2. Materials and Methods

2.1. Cell Culture

SH-SY5Y control cells, obtained from Dr. Kim H.J (KBRI), and the tandem mCherry–GFP tag fused to FIS1 stable SH-SY5Y cells, a kind gift from Dr. Ian G. Ganley (University of Dundee, Dundee, UK) [18], were grown in normal culture conditions with DMEM/F12 (ThermoFisher, Waltham, MA, USA) supplemented with 15% fetal bovine serum (ThermoFisher, USA), 100 units/mL of penicillin, and 100 ug/mL of streptomycin (ThermoFisher, USA) at 37 °C in a humidified 5% CO_2 atmosphere. The tandem mCherry–GFP tag fused to FIS1 stable SH-SY5Y cells were selected with 500 μg/mL of hygromycin (Sigma, St. Louis, MO, USA) and a stable pool was used for experiments.

2.2. Viability Assay

SH-SY5Y cells were plated and treated with PPA (0.1, 1, 2, 6, and 12 mM, Sigma, St. Louis, MO, USA) in 96-well plates for 48 and 72 h Approximately 10 µL of CCK-8 (Dojindo, Kumamoto, Japan) was added to cells, and the optical density (OD) value was measured at 450 nm.

2.3. Immunocytochemistry

SH-SY5Y cells were grown on coverslips and treated with 1 mM PPA (Sigma, USA) for 72 h Cells were fixed with 1% paraformaldehyde (EMS, Hatfield, PA, USA) in phosphate-buffered saline (PBS, Welgene, Gyeongsangbuk-do, Gyeongsan-si, Korea) containing 4% sucrose for 5 min at room temperature. Primary antibodies against LC3A/B (#12741, Cell Signaling, Danvers, MA, USA) were added with blocking solution containing 0.1% gelatin, 0.3% Triton X-100, 16 mM sodium phosphate, and 450 mM NaCl, and cells were incubated overnight at 4 °C. After being washed with PBS, coverslips were incubated with Alexa Fluor488 (#4412, Cell Signaling, USA)-conjugated secondary antibodies for 1 h at room temperature and then again washed with PBS. Next, coverslips were mounted with a mounting medium (H-1000, Vector Laboratories, Burlingame, CA, USA) and were imaged with fluorescence microscopy (Nikon, Tokyo, Japan) using a 488 nm fluorescence filter.

2.4. Transmission Electron Microscopy for Quantifying Autophagic Elements

SH-SY5Y cells were treated with 1 mM of PPA for 72 h and then fixed with 2.5% glutaraldehyde/2% paraformaldehyde solution for 2 h Fixed cells were then post-fixed with 2% osmium tetroxide (EMS, USA) for 2 h at 4 °C, and the block was stained in 0.1 mg thiocarbohydrazide (TCH, TCI, Tokyo, Japan) in 10 mL distilled water and 1% uranyl acetate (EMS, USA) and dehydrated with a graded ethanol series. The samples were then embedded with an EMBed-812 embedding kit (EMS, USA). The embedded samples were sectioned (60 nm) with an ultramicrotome (Leica, Wetzlar, Germany), and the sections were then viewed on a Tecnai G2 transmission electron microscope (Thermofisher) at 120 kV. The numbers of autophagosomes and autolysosomes per cell were assessed.

2.5. Correlative Light and Electron Microscopy

CLEM was performed as previously described [23]. The mCherry–GFP tag fused to FIS1 stable SH-SY5Y cells were grown in 35 mm glass grid-bottomed culture dishes to 50–60% confluency. Cells with or without 1 mM or 2 mM PPA treatment were stained with 100 nM LysoTracker (LysoTracker Blue DND-22, Thermofisher, USA) for 15 min and then imaged under a confocal light microscope (Ti-RCP, Nikon, Japan), and after 24 h treatment of PPA, cells were fixed with 1% glutaraldehyde and 1% paraformaldehyde in 0.1 M cacodylate solution (pH 7.0). After being washed, cells were dehydrated with a graded ethanol series and infiltrated with an embedding medium. After embedment, 60 nm sections were cut horizontally to the plane of the block (UC7; Leica Microsystems, Germany) and were mounted on copper slot grids with a specimen support film. Sections were stained with uranyl acetate and lead citrate. The cells were observed at 120 kV in a Tecnai G2 microscope (ThermoFisher, USA). Confocal micrographs were produced as high-quality large images using PhotoZoom Pro 8 software (Benvista Ltd., Houston, TX, USA). Enlarged fluorescence images were fitted to the electron micrographs using the Image J BigWarp program.

2.6. Measurement of Mitophagy

The mCherry–GFP tag fused to FIS1 stable SH-SY5Y cells (5×10^4 cells/well) was grown in 35 mm glass-bottomed culture dishes (MatTEK, Ashland, MA, USA) and treated with 1 and 2 mM PPA. Parallel incubation of cells without PPA was used for control. Measurement of mitophagy has been described previously [16]. Briefly, quantitation was performed for five fields of view for each group. Red-alone puncta were defined as round structures found only in the red channel with no corresponding structure in the green channel. Quantitative data were collected by manually counting all red-only puncta within each cell for each field of view.

3. Results

The viability of SH-SY5Y cells after treatment with PPA showed that optimal concentrations of PPA were less than 2 mM. Viability was assessed with CCK-8 assays after treatment with concentrations of 0, 0.1, 1, 2, 6, and 12 mM. Viability was significantly decreased in response to 2 mM after 48 h incubation with PPA (Figure 1A). After 72 h, treatment with 1 mM PPA a significant change was also shown (Figure 1B). Therefore, we used 1 mM PPA for 72 h to assess autophagy in SH-SY5Y cells. The number of LC3 puncta in PPA-treated cells was increased, as shown by immunofluorescence (Figure 1C). The number of LC3 puncta was 2.9 ± 0.28, compared with 1.8 ± 0.3293 for untreated cells (Figure 1D). No difference in the intensity of LC3 puncta was observed.

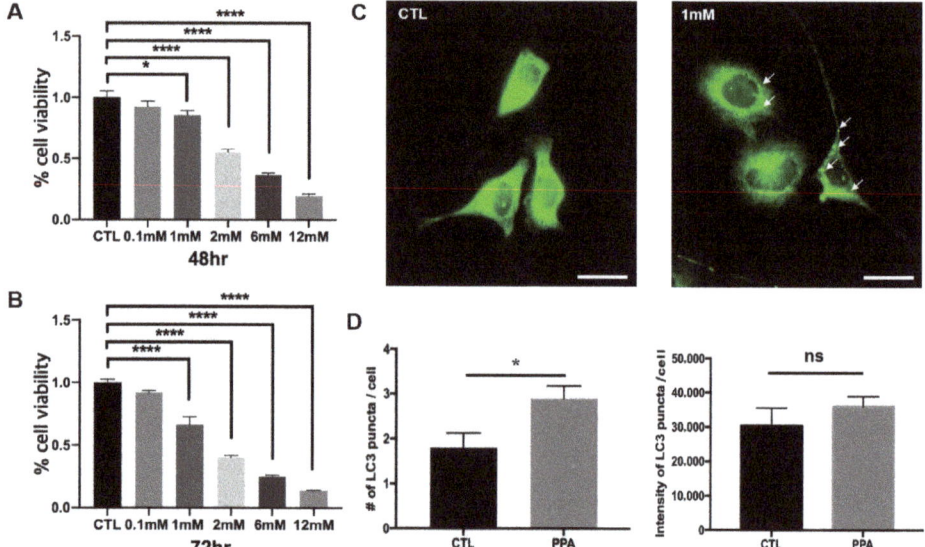

Figure 1. Viability assay and increase of of LC3 level in SH-SY5Y cells after propionic acid (PPA) treatment. The cells were treated with PPA for (**A**) 48 h and (**B**) 72 h Statistical analysis used a two- way ANOVA. Results are presented as mean ± SEM. When the concentration reached 2–12 mM, cell viability was significantly reduced. (**C**) Representative immunofluorescence images showing LC3 puncta in SH-SY5Y control cells and PPA-treated cells. The white scale bar is 50 μm. (**D**) Numbers of LC3 puncta and LC3 puncta intensity from images in (**A**) (n = 5), illustrating a significant increase of number of LC3 puncta following treatment with 1 mM PPA. Statistical analysis used a one-way ANOVA. Results are presented as mean ± SEM. * $p < 0.05$, **** $p < 0.0001$.

We further analyzed the number of autophagosomes and autolysosomes in control and PPA-treated cells using TEM images. The numbers of both organelles were increased in PPA-treated cells. The numbers of autophagosomes per cell were 6.1 ± 0.7371 and 1.9 ± 0.4333 for treated and control cells, respectively (Figure 2A). Similar results for autolysosomes were found: 6.6 ± 1.067 for PPA-treated cells and 1.9 ± 0.5667 for untreated cells (Figure 2B).

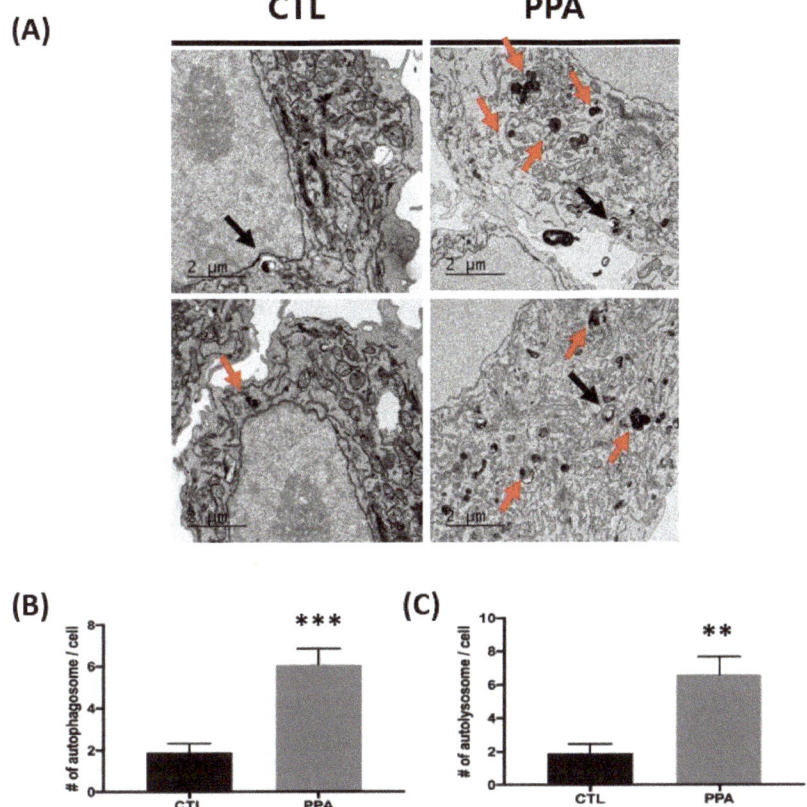

Figure 2. The increase in autophagy following PPA treatment in SH-SY5Y cells using TEM. (**A**) Representative TEM images showing autophagy in control SH-SY5Y cells and PPA-treated cells. The black arrow indicates autophagosomes, and red arrows indicate autolysosomes. (**B** and **C**) Numbers of autophagosomes and autolysosomes from images in (**A**) (n = 10). TEM analysis shows that numbers of both autophagosomes and autolysosomes increase in cells treated with 1 mM PPA. Statistical analysis used a one-way ANOVA. Results are presented as mean ± SEM. ** $p < 0.01$, *** $p < 0.0005$.

PPA was reported as a small molecule leading to mitochondrial dysfunction [7]. Therefore, we used the tandem mCherry–GFP tag fused to FIS1-stable SH-SY5Y cells to confirm the induction of mitophagy by PPA treatment. GFP and mCherry show green and red fluorescence, respectively, with the former specific for mitochondria and the latter for mitophagy (Figure 3A). After treatment with 1 or 2 mM PPA, the number of mCherry red puncta increased 4.6 times, indicating the induction of mitophagy in treated cells (Figure 3B).

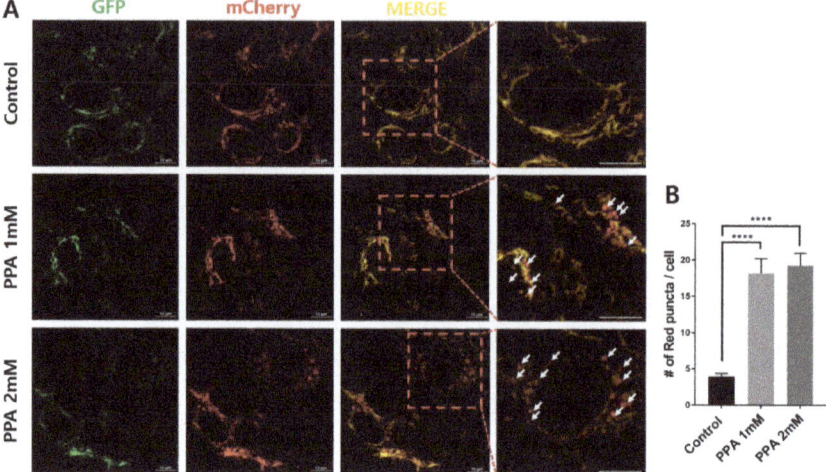

Figure 3. Mitophagy assays in SH-SY5Y cells following treatment with 1 and 2 mM PPA for 24 h. (**A**) mCherry–GFP-FIS1101–152 stably expressing cells were subjected to (1) control, (2) 1 mM PPA, and (3) 2 mM PPA for 24 h. (**B**) Numbers of red puncta (mitophagy) per cell (n > 40). Statistical analysis used a one-way ANOVA. Results are presented as mean ± SEM. **** $p < 0.0001$.

We employed CLEM to confirm the ultrastructure of red puncta (Figure 4). Live cells were imaged using confocal microscopy treatment with PPA after 24 h. Images were then aligned with stitched TEM images of the same cells. Almost healthy mitochondria showed the green fluorescence of GFP, and some red puncta were co-localized with LysoTracker (Figure 4A). In control cells, LysoTracker-positive structures (blue) are seen contacting red puncta (Figure 4A, white dot line box, and enlarged images). In the PPA treatment cells, damaged round mitochondria show high electron density in the electron micrographs and are co-localized with LysoTracker, suggesting mitophagy (Figure 4B).

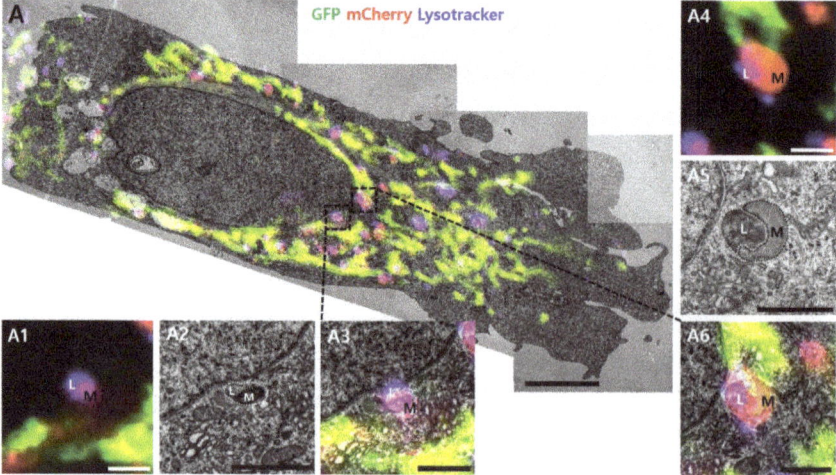

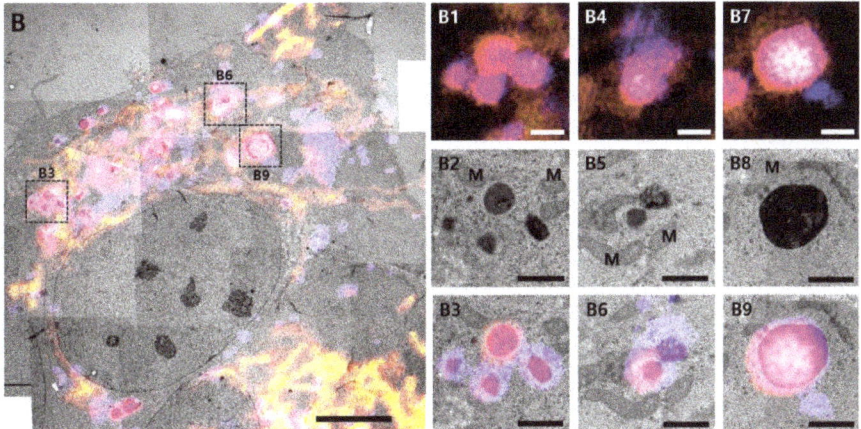

Figure 4. Correlative light and electron microscopy. Correlative confocal and electron microscope images of (**A**) co-localized mCherry–GFP-FIS1101–152 in SH-SY5Y cells or (**B**) cells treated with 2 mM PPA. Yellow indicates co-localization of GFP and mCherry signals. Magenta indicates co-localization of mCherry and LysoTracker. Multiple TEM images were taken at 1700× magnification. Images were stitched for a large field of view at higher resolution. The black dot line box indicates structures corresponding to magenta fluorescent puncta on fluorescence images. The structures showed the morphology of mitophagy, as demonstrated by the black dot line box shown at higher magnification in the inserted images (A1–A6 and B1–B9). L, lysosome; M, mitochondria. Size bar in A and B = 5 µm, A1~A6 = 1 µm, B1~B9 = 1 µm.

4. Discussion

Mitochondria are continuously replenished. As new mitochondria are produced, dysfunctional organelles are removed by autophagy-mediated degradation though mitophagy [24]. There are several control or repair systems for mitochondrial structure and function maintaining essential energy metabolism. Oxidatively damaged proteins in the mitochondrial outer membrane can be degraded by the ubiquitin–proteasome system. When damage is more extensive, e.g., through exposure to elevated reactive oxygen species (ROS) or aging, mitochondria are sequestered by autophagosome and fused with lysosome for degradation. It is called mitophagy. The extent of mitophagy in neurites is influenced by various factors related to mitochondrial function, and the contribution of mitophagy to mitochondrial function in soma or neurites is critical to understanding the regulation of energy metabolism in these cells [25]. Our previous work shows that PPA induces defects in mitochondria [6] and autophagy [10]. In this study, we investigated the relationship between mitochondrial dysfunction and increased autophagy.

Some mitochondrial toxins, including rotenone, concomitantly activate autophagy, including mitophagy. Like rotenone, PPA-treated cells showed elevated autophagic sequestration of mitochondria. More prominent LC3 signals (Figure 1C) and LC3-II/β-actin ratios (data not shown) indicate change of autophagy in PPA-treated cells. Several reports that focus on mitochondrial function in PPA-treated cells are available. Kim et al. showed an increase of mitochondrial copy number and expression of PGC-1a, COX4, SIRT3, and, TFAM (mitochondrial biogenesis-related proteins) after PPA treatment of SH-SY5Y cells [7]. Dysfunction of mitochondria caused by RNA interference-mediated knockdown of peroxisome proliferator-activated receptor-γ coactivator 1α (PGC-1α) in neurons showed abnormal synapse formation in developing neural circuits and failure to maintain synapses in the hippocampus of adults [26]. Induction of mitochondrial biogenesis following expression of PGC-1α is stimulated by brain-derived neurotrophic factor, which can be modulated by changes of SCFAs in the brain [27]. PPA-induced mitochondrial dysfunction suggests a mechanism for neurotoxicity. El-Ansary et al. showed PPA-induced neurotoxicity in rat pups though depletion of gamma-aminobutyric acid and serotonin. [28] Frye et al. showed oxidative stress after PPA exposure, and Alfawaz et al. showed that factors related to mitochondria, such as carnosine, N-acetylcysteine, and vitamin D, can rescue neurotoxicity caused by PPA in rat pups [29]. The protective effect of carnosine (β-alanyl-L-histidine) is related to autophagy and causes a decrease in Drp-1 expression. Further, treatment with N-acetylcysteine shows inhibition of Atg32-dependent mitophagy [29]. In PPA-treated SH-SY5Y cells in the present study, TEM analysis shows an increase in numbers of both autophagosomes and autolysosomes, which reflects properly functioning autophagy flux (Figure 2).

Several technical challenges using fluorescence and biochemical techniques to analyze autophagic processes, including mitophagy, are recognized [17,30]. Such challenges were met in this study using CLEM techniques to monitor autophagy for cellular homeostasis due to mitochondrial dysfunction in PPA-treated SH-SY5Y cells. The tandem mCherry–GFP tag fused to FIS1 was used in our approach. Red fluorescence of mCherry increased in PPA-treated cells, suggesting increased mitophagy, and the green fluorescent of GFP in mitochondria did not change significantly (Figure 3). CLEM confirmed the ultrastructure associated with these colors as mitochondria and mitophagy (Figure 4). There is a mismatch between some healthy mitochondrial fluorescence signal (yellow color) and EM images due to technical limitations of the TEM based CLEM method (Supplementary Figure S1 A1 and B1). It is difficult to accurately match the Z axis of the optical section (LM) and the physical section (EM), since LM and EM image thicknesses are different. Due to the difference, there is more information in the fluorescence micrograph (LM: 300 nm optical thickness, EM: 60 nm physical thickness). Although there is some technical limitation, the damaged mitochondria are well correlated with the lysotracker (Supplementary Figure S1, black arrow). Thus, observations depicted in Figures 3 and 4 indicate that increased autophagy shown in Figures 1 and 2 is mitophagy.

5. Conclusions

Changes in mitophagy under stress condition is associated with pathological conditions, including neurodegenerative diseases and myopathies. Therefore, identifying mitophagy modulators and understanding their mechanisms of action will provide critical insight into neurodegenerative diseases. We confirmed that mitophagy was induced by PPA treatment in SH-SY5Y cells. CLEM is a useful technique for monitoring mitophagy in cells under stress. Various stages of mitophagy, including initiation of autophagy, vesicle completion, lysosome fusion, and degradation of mitochondria in lysosomes, can be monitored in CLEM, if the time points in such studies are adequately controlled. CLEM might also be applied to study structural changes of other cellular organelles.

Supplementary Materials: The following are available online at http://www.mdpi.com/1996-1944/13/19/4336/s1, Figure S1: Correlative confocal and electron microscope images of the control (A) and 2 mM PPA treated cell (B).

Author Contributions: Conceptualization, M.J., H.C., and J.Y.M.; methodology, M.J., H.C., and; formal analysis, M.J., H.C.; investigation, M.J., H.C., and J.Y.M.; resources, J.K.; data curation, M.J., H.C.; writing—original draft preparation, J.Y.M.; writing—review and editing, M.J., H.C., J.K., and J.Y.M.; visualization, M.J., H.C.; supervision, J.Y.M.; project administration, J.Y.M.; funding acquisition, J.K. and J.Y.M. All authors have read and agreed to the published version of the manuscript.

Funding: This research was supported by the National Research Foundation of Korea (NRF) grant funded by the government of Korea (MSIP) (No. 2019R1A2C1010634), the Organelle Network Research Center (NRF-2017R1A5A1015366), and KBRI basic research program though the Korea Brain Research Institute funded by the Ministry of Science and ICT (20-BR-02-09).

Acknowledgments: Instrument (transmission electron microscopy and confocal microscopy) data were acquired at the Brain Research Core Facilities in KBRI.

Conflicts of Interest: The authors declare no conflict of interest.

References

1. Den Besten, G.; van Eunen, K.; Groen, A.K.; Venema, K.; Reijngoud, D.J.; Bakker, B.M. The role of short-chain fatty acids in the interplay between diet, gut microbiota, and host energy metabolism. *J. Lipid Res.* **2013**, *54*, 2325–2340. [CrossRef] [PubMed]
2. Foley, K.A.; MacFabe, D.F.; Vaz, A.; Ossenkopp, K.P.; Kavaliers, M. Sexually dimorphic effects of prenatal exposure to propionic acid and lipopolysaccharide on social behavior in neonatal, adolescent, and adult rats: Implications for autism spectrum disorders. *Int. J. Dev. Neurosci.* **2014**, *39*, 68–78. [CrossRef]
3. El-Ansary, A.; Al-Ayadhi, L. Relative abundance of short chain and polyunsaturated fatty acids in propionic acid-induced autistic features in rat pups as potential markers in autism. *Lipids Health Dis.* **2014**, *13*, 140. [CrossRef] [PubMed]
4. Nankova, B.B.; Agarwal, R.; MacFabe, D.F.; La Gamma, E.F. Enteric bacterial metabolites propionic and butyric acid modulate gene expression, including CREB-dependent catecholaminergic neurotransmission, in PC12 cells–possible relevance to autism spectrum disorders. *PLoS ONE* **2014**, *9*, e103740. [CrossRef]
5. MacFabe, D.F.; Cain, N.E.; Boon, F.; Ossenkopp, K.P.; Cain, D.P. Effects of the enteric bacterial metabolic product propionic acid on object-directed behavior, social behavior, cognition, and neuroinflammation in adolescent rats: Relevance to autism spectrum disorder. *Behav. Brain Res.* **2011**, *217*, 47–54. [CrossRef] [PubMed]
6. Duscha, A.; Gisevius, B.; Hirschberg, S.; Yissachar, N.; Stangl, G.I.; Eilers, E.; Bader, V.; Haase, S.; Kaisler, J.; David, C.; et al. Propionic Acid Shapes the Multiple Sclerosis Disease Course by an Immunomodulatory Mechanism. *Cell* **2020**, *180*, 1067–1080.e16. [CrossRef]
7. Kim, S.A.; Jang, E.H.; Mun, J.Y.; Choi, H. Propionic acid induces mitochondrial dysfunction and affects gene expression for mitochondria biogenesis and neuronal differentiation in SH-SY5Y cell line. *Neurotoxicology* **2019**, *75*, 116–122. [CrossRef]
8. Frye, R.E.; Melnyk, S.; Macfabe, D.F. Unique acyl-carnitine profiles are potential biomarkers for acquired mitochondrial disease in autism spectrum disorder. *Transl. Psychiatry* **2013**, *3*, e220. [CrossRef]
9. Tang, Y.; Chen, Y.; Jiang, H.; Nie, D. Short-chain fatty acids induced autophagy serves as an adaptive strategy for retarding mitochondria-mediated apoptotic cell death. *Cell Death Differ.* **2011**, *18*, 602–618. [CrossRef]
10. Choi, H.; Kim, I.S.; Mun, J.Y. Propionic acid induces dendritic spine loss by MAPK/ERK signaling and dysregulation of autophagic flux. *Mol. Brain* **2020**, *13*, 86. [CrossRef]
11. Fivenson, E.M.; Lautrup, S.; Sun, N.; Scheibye-Knudsen, M.; Stevnsner, T.; Nilsen, H.; Bohr, V.A.; Fang, E.F. Mitophagy in neurodegeneration and aging. *Neurochem. Int.* **2017**, *109*, 202–209. [CrossRef] [PubMed]
12. Wang, Y.; Liu, N.; Lu, B. Mechanisms and roles of mitophagy in neurodegenerative diseases. *CNS Neurosci. Ther.* **2019**, *25*, 859–875. [CrossRef] [PubMed]
13. Nixon, R.A. The role of autophagy in neurodegenerative disease. *Nat. Med.* **2013**, *19*, 983–997. [CrossRef] [PubMed]
14. Chakraborty, J.; Caicci, F.; Roy, M.; Ziviani, E. Investigating mitochondrial autophagy by routine transmission electron microscopy: Seeing is believing? *Pharmacol. Res.* **2020**, *160*, 105097. [CrossRef]
15. Jung, M.; Choi, H.; Mun, J.Y. The autophagy research in electron microscopy. *Appl. Microsc.* **2019**, *49*, 11. [CrossRef]
16. Katayama, H.; Kogure, T.; Mizushima, N.; Yoshimori, T.; Miyawaki, A. A sensitive and quantitative technique for detecting autophagic events based on lysosomal delivery. *Chem. Biol.* **2011**, *18*, 1042–1052. [CrossRef]
17. Sun, N.; Malide, D.; Liu, J.; Rovira, I.I.; Combs, C.A.; Finkel, T. A fluorescence-based imaging method to measure in vitro and in vivo mitophagy using mt-Keima. *Nat. Protoc.* **2017**, *12*, 1576–1587. [CrossRef]
18. Allen, G.F.; Toth, R.; James, J.; Ganley, I.G. Loss of iron triggers PINK1/Parkin-independent mitophagy. *EMBO Rep.* **2013**, *14*, 1127–1135. [CrossRef]
19. Swanlund, J.M.; Kregel, K.C.; Oberley, T.D. Investigating autophagy: Quantitative morphometric analysis using electron microscopy. *Autophagy* **2010**, *6*, 270–277. [CrossRef]
20. Mizushima, N.; Yoshimori, T.; Levine, B. Methods in mammalian autophagy research. *Cell* **2010**, *140*, 313–326. [CrossRef]
21. Gudmundsson, S.; Kahlhofer, J.; Baylac, N.; Kallio, K.; Eskelinen, E.L. Correlative Light and Electron Microscopy of Autophagosomes. *Methods Mol. Biol.* **2019**, *1880*, 199–209. [PubMed]

22. Chen, L.; Ma, K.; Han, J.; Chen, Q.; Zhu, Y. Monitoring Mitophagy in Mammalian Cells. *Methods Enzymol.* **2017**, *588*, 187–208.
23. Reddick, L.E.; Alto, N.M. Correlative light and electron microscopy (CLEM) as a tool to visualize microinjected molecules and their eukaryotic sub-cellular targets. *J. Vis. Exp.* **2012**, *63*, e3650. [CrossRef] [PubMed]
24. Evans, C.S.; Holzbaur, E.L.F. Quality Control in Neurons: Mitophagy and Other Selective Autophagy Mechanisms. *J. Mol. Biol.* **2020**, *432*, 240–260. [CrossRef] [PubMed]
25. Martinez-Vicente, M. Neuronal Mitophagy in Neurodegenerative Diseases. *Front. Mol. Neurosci.* **2017**, *10*, 64. [CrossRef]
26. Cheng, A.; Wan, R.; Yang, J.L.; Kamimura, N.; Son, T.G.; Ouyang, X.; Luo, Y.; Okun, E.; Mattson, M.P. Involvement of PGC-1alpha in the formation and maintenance of neuronal dendritic spines. *Nat. Commun.* **2012**, *3*, 1250. [CrossRef]
27. Maqsood, R.; Stone, T.W. The Gut-Brain Axis, BDNF, NMDA and CNS Disorders. *Neurochem. Res.* **2016**, *41*, 2819–2835. [CrossRef]
28. El-Ansary, A.K.; Al-Daihan, S.K.; El-Gezeery, A.R. On the protective effect of omega-3 against propionic acid-induced neurotoxicity in rat pups. *Lipids Health Dis.* **2011**, *10*, 142. [CrossRef]
29. Alfawaz, H.A.; Bhat, R.S.; Al-Ayadhi, L.; El-Ansary, A.K. Protective and restorative potency of Vitamin D on persistent biochemical autistic features induced in propionic acid-intoxicated rat pups. *BMC Complement. Altern Med.* **2014**, *14*, 416. [CrossRef]
30. Mauro-Lizcano, M.; Esteban-Martinez, L.; Seco, E.; Serrano-Puebla, A.; Garcia-Ledo, L.; Figueiredo-Pereira, C.; Vieira, H.L.; Boya, P. New method to assess mitophagy flux by flow cytometry. *Autophagy* **2015**, *11*, 833–843. [CrossRef]

© 2020 by the authors. Licensee MDPI, Basel, Switzerland. This article is an open access article distributed under the terms and conditions of the Creative Commons Attribution (CC BY) license (http://creativecommons.org/licenses/by/4.0/).

Article

Exploiting the Acceleration Voltage Dependence of EMCD

Stefan Löffler [1,*], Michael Stöger-Pollach [1], Andreas Steiger-Thirsfeld [1], Walid Hetaba [2] and Peter Schattschneider [1,3]

1 University Service Centre for Transmission Electron Microscopy, TU Wien,
Wiedner Hauptstraße 8-10/E057-02, 1040 Wien, Austria; michael.stoeger-pollach@tuwien.ac.at (M.S.-P.); andreas.steiger-thirsfeld@tuwien.ac.at (A.S.-T.); peter.schattschneider@tuwien.ac.at (P.S.)
2 Max Planck Institute for Chemical Energy Conversion, Stiftstraße 34-36, 45470 Mülheim an der Ruhr, Germany; hetaba@fhi-berlin.mpg.de
3 Institute of Solid State Physics, TU Wien, Wiedner Hauptstraße 8-10/E138-03, 1040 Wien, Austria
* Correspondence: stefan.loeffler@tuwien.ac.at

Abstract: Energy-loss magnetic chiral dichroism (EMCD) is a versatile method for measuring magnetism down to the atomic scale in transmission electron microscopy (TEM). As the magnetic signal is encoded in the phase of the electron wave, any process distorting this characteristic phase is detrimental for EMCD. For example, elastic scattering gives rise to a complex thickness dependence of the signal. Since the details of elastic scattering depend on the electron's energy, EMCD strongly depends on the acceleration voltage. Here, we quantitatively investigate this dependence in detail, using a combination of theory, numerical simulations, and experimental data. Our formulas enable scientists to optimize the acceleration voltage when performing EMCD experiments.

Keywords: EMCD; TEM; EELS; magnetism; acceleration voltage

1. Introduction

Circular dichroism in X-ray Absorption Spectroscopy (XAS) probes the chirality of the scatterer, related either to a helical arrangement of atoms or to spin polarized transitions as studied in X-ray Magnetic Circular Dichroism (XMCD). Before the new millenium, it was considered impossible to see such chirality in electron energy-loss spectrometry (EELS). On the other hand, the formal equivalence between the polarization vector in XAS and the scattering vector in EELS tells us that any effect observable in XAS should have its counterpart in EELS. For instance, anisotropy in XAS corresponds to anisotropy of the double differential scattering cross section (DDSCS) in EELS. A well known example is the directional prevalence of either $s \to \pi^*$ and $s \to \sigma^*$ transitions in the carbon K-edge of graphite, depending on the direction of the scattering vector [1,2].

In XMCD, the polarization vector is helical—a superposition of two linear polarization vectors $\mathbf{e}_x \pm i\mathbf{e}_y$ orthogonal to each other—resembling a left- and right-handed helical photon, respectively. However, what is the counterpart of photon helicity in EELS?

In 2002, one of the authors and their postdoc speculated about what the counterpart of photon helicity could be in EELS—an arcane issue at the time. This led to a keen proposal to study spin polarized transitions in the electron microscope [3]. Closer inspection revealed that in EELS, a superposition of two scattering vectors orthogonal to each other with a relative phase shift of $\pm \pi/2$ is needed, exactly as the formal similarity with XMCD dictated. This, in turn, called for a scattering geometry that exploits the coherence terms in the DDSCS [4,5]. These insights led to the CHIRALTEM project [6].

The multidisciplinary team elaborated the appropriate geometry for the analysis of ionization edges in the spirit of XMCD. The first EELS spectrum was published in 2006 [7]. In that paper, the new method was baptized EMCD—Electron (Energy Loss) Magnetic Chiral Dichroism—in analogy to XMCD. The term "chiral" was deliberately chosen instead of "circular" because the chirality of electronic transitions was to be detected, and because there is no circular polarization in EELS. The experiment confirmed that the physics behind EMCD is very similar to the physics of XMCD. Rapid progress followed: consolidation of the theory [8,9], optimization of experimental parameters [10], dedicated simulation software [11,12], and spatial resolution approaching the nm [13,14] and the atomic scale [15–23].

A genuine feature of EMCD is the ability to probe selected crystallographic sites [18,24], e.g., in Heusler alloys [25], ferrimagnetic spinels [26], or perovskites [27,28]. The high spatial resolution of the method allows the study of nanoparticles [14], 3d–4f coupling in superlattices [29], specimens with stochastically oriented crystallites and even of amorphous materials [30]. EMCD has also been used to investigate spin polarization of non-magnetic atoms in dilute magnetic semiconductors [31], magnetic order breakdown in MnAs [32], GMR of mixed phases [33] and magnetotactic bacteria [34]. A key experiment on magnetite, exploiting the combination of atomic resolution in STEM with the site specificity showed the antiferromagnetic coupling of adjacent Fe atoms directly in real space [16]. An overview of EMCD treating many aspects of anisotropy and chirality in EELS can be found in [35].

To date, EMCD measurements have predominantly been performed at the highest available acceleration voltages—typically 200 keV to 300 keV—which has several advantages such as better resolution, a larger inelastic mean free path, and optimal detector performance resulting in a reasonable signal-to-noise ratio. However, by limiting oneself to a specific acceleration voltage and hence electron energy, EMCD cannot be used to its full potential.

One example where choosing a lower acceleration voltage can be tremendously helpful is the reduction or avoidance of beam damage [36–39]. Another is the investigation of the magnetization dependence: in a TEM, the sample is placed inside the objective lens with a typical field strength of the order of 2 T for 200 keV electrons. By changing the acceleration voltage, the objective lens field applied at the sample position is changed as well [40], thereby enabling magnetization-dependent investigations. This can even be used to drive magnetic field induced phase transitions [27]. Moreover, EMCD is strongly affected by elastic scattering, and, hence, thickness and sample orientation [8,11,25,41]. Therefore, changing the electron energy and therefore the details of the elastic scattering processes enables EMCD measurements even at a thickness and orientation where no significant EMCD effect is observable at a high acceleration voltage. This proposition is corroborated by early numerical simulations [42], which to our knowledge have not been followed up on or widely adopted by the community.

2. Results
2.1. Theory

The general formula governing EMCD has already been outlined in the original publications theoretically predicting the effect and demonstrating it experimentally [3,7]. Detailed ab initio studies soon followed [8]. However, those formulations all aimed at very high accuracy; none of them gave a simple, closed form to quickly calculate the EMCD effect and easily see the influence parameters such as, e.g., the acceleration voltage have on the outcome. Recently, Schneider et al. [41] published such a formula; however, they neglected any elastic scattering the beam can undergo after an inelastic scattering event by approximating the outgoing wave by a simple plane wave.

Here, we present a derivation of a simple formula taking into account elastic scattering both before and after the inelastic scattering event. In the process, we will make four major assumptions:

1. We limit the derivation to an incident three-beam and outgoing two-beam case in the zero-order Laue zone of a sample that is single-crystalline in the probed region with a centro-symmetric crystal structure;
2. We assume that the sample is a slab of thickness t with an entrance and an exit plane essentially perpendicular to the beam propagation axis;
3. We assume that the inelastic scattering process is at least four-fold rotationally symmetric around the optical axis and that the characteristic momentum transfer q_e is

much smaller than the chosen reciprocal lattice distance $|G|$. This implies that the inelastic scattering in the chosen geometry is only dependent on the scattering atom's spin-state, but not influenced significantly by any anisotropic crystal field;

4. We assume that the atoms of the investigated species are homogeneously distributed along the beam propagation axis and that $G \cdot x = 2m\pi, m \in \mathbb{Z}$ for all atom positions x and the chosen lattice vector G.

Assumption 1 comes from the conventional EMCD setup: the (crystalline) sample is tilted into systematic row condition and the detector is placed on (or close to) the Thales circle between neighboring diffraction spots. In a symmetric systematic row condition, the strongest diffraction spots are the central one (0) and the two diffraction spots at $-G, G$, which have the same intensity. Any higher-order diffraction spots are comparatively weak and will therefore be neglected.

To understand the reason behind the outgoing two-beam case, we follow the reciprocity theorem [43,44]. A (point-like) detector in reciprocal space detects exact plane-wave components. If we trace those back to the exit plane of the sample, we can expand them into Bloch waves. For the typical EMCD detector positions, they correspond exactly to the Bloch waves we get in a two-beam case (where the Laue circle center is positioned somewhere along the bisector of the line from 0 to G).

The probability of measuring a particular state $|\psi_{\text{out}}\rangle$ (a "click" in the detector corresponding to a plane wave at the exit plane of the sample) given a certain incident state $|\psi_{\text{in}}\rangle$ (a plane wave incident on the entry plane of the sample) is given by Fermi's Golden rule [45–49]:

$$p = \sum_{I,F} p_I (1 - p_F) \langle \psi_{\text{out}} | \langle F | \hat{V} | I \rangle | \psi_{\text{in}} \rangle \langle \psi_{\text{in}} | \langle I | \hat{V}^\dagger | F \rangle | \psi_{\text{out}} \rangle \delta(E_F - E_I - E), \quad (1)$$

where I, F run over all initial and final states of the sample, p_I, p_F are their respective occupation probabilities, E_I, E_F are their respective energies, E is the EELS energy loss, and $\hat{V}$ is the transition operator. In momentum representation, $\hat{V}$ for a single atom is given by

$$\langle \tilde{k} | \hat{V} | k \rangle = \frac{e^{iq \cdot \hat{R}}}{q^2} \quad \text{with} \quad q = k - \tilde{k}. \quad (2)$$

With the mixed dynamic form factor (MDFF) [45,49–51],

$$S(q, q', E) = \sum_{I,F} p_I (1 - p_F) \langle \tilde{k} | \langle F | e^{iq \cdot \hat{R}} | I \rangle | k \rangle \langle k' | \langle I | e^{-iq' \cdot \hat{R}} | F \rangle | \tilde{k}' \rangle \delta(E_F - E_I - E), \quad (3)$$

the probability for a "click" in the detector can be written as [8,45,48–50]

$$p = \iiiint \sum_x e^{i(q-q') \cdot x} \psi_{\text{out}}(\tilde{k})^* \psi_{\text{out}}(\tilde{k}') \frac{S(q, q', E)}{q^2 q'^2} \psi_{\text{in}}(k) \psi_{\text{in}}(k')^* dk dk' d\tilde{k} d\tilde{k}', \quad (4)$$

where the $\sum_x e^{i(q-q') \cdot x}$ stems from the summation over all atoms (of the investigated species) in the sample and the MDFF is taken to be the MDFF of a single such atom located at the origin.

Specific expressions for the MDFF for various models under different conditions and approximations are well known (see, e.g., [7,49,52]), but their details will be irrelevant for the majority of our derivation for which we will keep the general expression $S(q, q', E)$.

Using the Bloch wave formalism [8,36,53–55], the three-beam incident wavefunction and the two-beam outgoing wave function can be written as

$$|\psi_{\text{in}}\rangle = \sum_{j \in \{1,2,3\}} \sum_{g \in \{-G,0,G\}} C^*_{j,0} C_{j,g} |\chi + \gamma_j n + g\rangle \quad (5)$$

$$|\psi_{\text{out}}\rangle = \sum_{l \in \{1,2\}} \sum_{h \in \{0,G\}} \tilde{C}^*_{l,0} e^{-i\tilde{\gamma}_l t} \tilde{C}_{l,h} |\tilde{\chi} + \tilde{\gamma}_l \tilde{n} + h\rangle, \quad (6)$$

where j, l are the Bloch wave indices, g, h run over the diffraction spots, the $C_{j,g}$ are the Bloch wave coefficients, the γ_j are the so-called anpassung, n is the surface normal vector, t is the sample thickness, and $\chi, \tilde{\chi}$ are the wave vectors of the incident and outgoing plane waves, respectively.

The derivation of the EMCD effect can be found in Appendix A. The final expression is

$$\eta = \frac{A \sin^2(\kappa t) - B \sin^2(\kappa' t)}{t + C \sin(2\kappa t)} \cdot \frac{\Im[S(q_1, q_2, E)]}{S(q_1, q_1, E)}, \quad (7)$$

where t is the sample thickness and the coefficients A, B, C, κ, κ' are defined in Equation (A18) (with Equations (A1) and (A3)).

Figure 1 shows a comparison of the thickness dependence predicted by Equation (7) and a full simulation based on Equation (4) for some typical, simple magnetic samples. Owing to the approximations made in the derivation, there naturally are some small differences (which are more pronounced at small thicknesses), but they are well within typical experimental uncertainties.

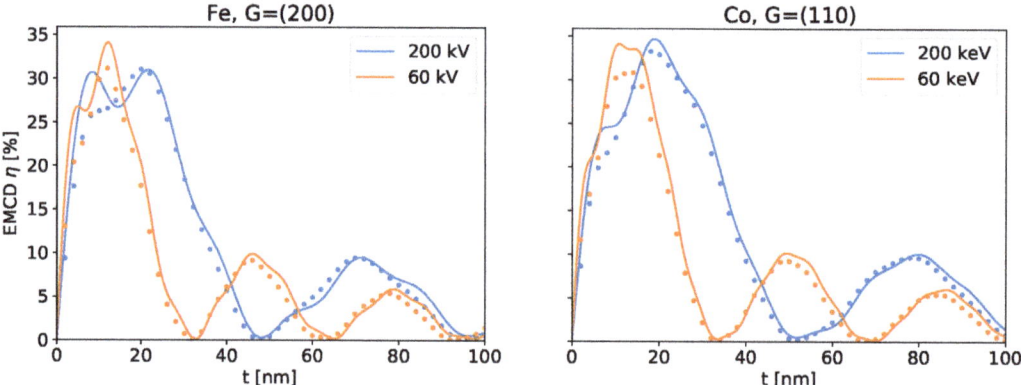

Figure 1. Comparison of the thickness dependence of the EMCD effect η predicted by Equation (7) (solid lines) and by the "bw" software using Equation (4) (dotted lines) for different acceleration voltages for bcc Fe and hcp Co.

Two main conclusions about the thickness-variation of the EMCD effect can be drawn from Equation (7). On the one hand, the numerator nicely shows the oscillatory nature of the effect. On the other hand, the denominantor clearly implies that the strength of the EMCD effect decreases approximately as $1/t$.

The numerator is composed of two oscillations with different amplitudes (A, B) and the frequencies

$$\kappa = \frac{\gamma_1 - \gamma_2}{2} = \frac{\sqrt{(|G|^2 - U_{2G})^2 + 8U_G^2}}{4\chi \cdot n} \quad \text{and} \quad \kappa' = \frac{\tilde{\gamma}_1 - \tilde{\gamma}_2}{2} = \frac{U_G}{2\tilde{\chi} \cdot \tilde{n}}, \quad (8)$$

which are closely related to the extinction distances for the incident and outgoing beams. As the wavevectors $\chi, \tilde{\chi}$ scale with the square root of the acceleration voltage $\sqrt{V}$, the frequencies of the oscillations of the EMCD effect scale with $1/\sqrt{V}$. This is corroborated by Figure 2.

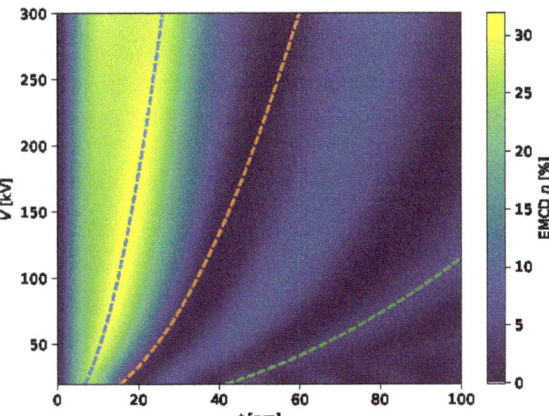

Figure 2. EMCD effect η for various acceleration voltages V and thicknesses t for bcc Fe as simulated with "bw". The dashed lines show (arbitrary) curves with $t \propto \sqrt{V}$ as guides for the eye.

Both the oscillations and the $1/t$ decay can be understood from the fact that EMCD is essentially an interferometry experiment. As such, it crucially depends on the relative phases of the different density matrix components after traversing the sample from the scattering center to the exit plane. Some scattering centers are positioned in a way that the resulting components contribute positively to the EMCD effect, other scattering centers are positioned such that their contribution to the EMCD effect is negative. As a result, there are alternating "bands" of atoms contributing positively and negatively [11], where the size of the bands is related to the extinction length. With increasing thickness, more and more alternating bands appear—the non-magnetic signal increases linearly with t, but the magnetic EMCD signal of all but one band averages out, ultimately resulting in a $1/t$ behavior of the relative EMCD effect.

Our theoretical results have several important implications. First, the EMCD effect can indeed be recorded at a wide variety of acceleration voltages as already proposed on numerical grounds in [42], thereby enabling magnetization-dependent measurements. Second, the thickness dependence scales with $1/t$, thus necessitating thin samples. Third, for a given sample thickness in the region of interest, a candidate for the optimal high tension yielding the maximal EMCD effect can easily be identified based on any existing simulation and the $\sqrt{V}$ scaling behavior (note, however, that other effects such as multiple plasmon scattering can put further constraints on the useful range of sample thicknesses, particularly at very low voltages).

2.2. Experiments

To corroborate our theoretical finding, we performed experiments at various high tensions to compare to the simulations. The experiments were performed on a ferrimagnetic magnetite (Fe_3O_4) sample [56], which has the advantage over pure Fe that it is unaffected by oxidation (it may, however, be partially reduced to Wüstite by prolonged ion or electron irradiation). The individual recorded spectra are shown in Figure 3. It is clearly visible that the EMCD effect changes with the high tension as predicted in Section 2.1. A quantitative comparison between the calculations and the experiments is shown in Figure 4 and shows excellent agreement.

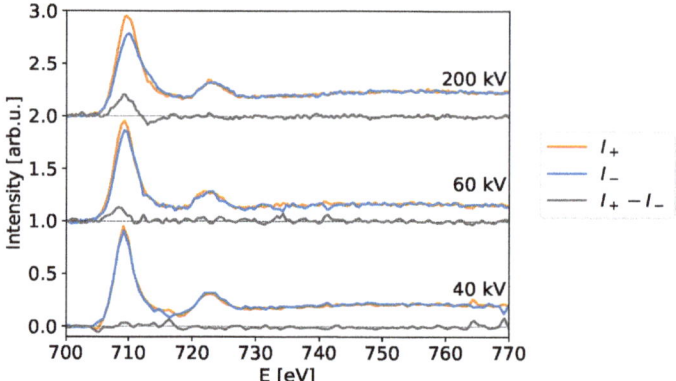

Figure 3. EMCD spectra for different acceleration voltages (as indicated) after background subtraction and post-edge normalization using the Fe L-edge in Magnetite tilted to a (4 0 0) systematic row condition. The sample-thickness was determined to be $t \approx 35$ nm for the 40 kV and 60 kV measurement positions and $t \approx 45$ nm for the 200 kV measurement position.

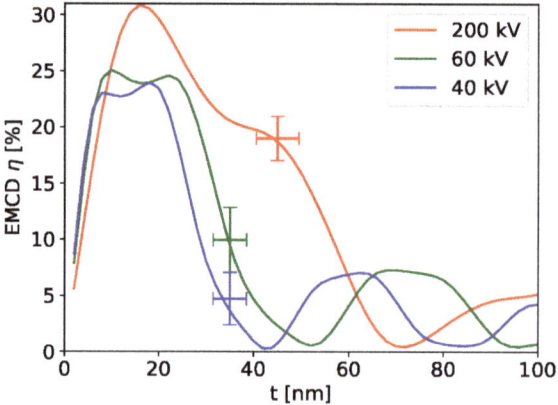

Figure 4. Comparison between numerical EMCD simulations ("bw", solid curves) and experiments (points) for Magnetite for three different acceleration voltages. For the experimental points, η was calculated from the data in Figure 3 according to Equation (9), the measured thickness values are given in the caption of Figure 3, and the error bars were determined as described in [57,58].

3. Discussion

Although Equation (7) is—to our knowledge—the first complete, analytical, closed form predicting the EMCD effect, several assumptions and approximations were made in its derivation. As such it is no replacement for full simulations with sophisticated software packages if ultimate accuracy is vital. Nevertheless, it can be a good starting point for EMCD investigations, and it helps elucidating the underlying physical principles and understanding the effects the experimental parameters have on EMCD. In this section, we will discuss the limits of the theoretical derivation based on the approximations made.

Assumption one deals with the scattering geometry and the crystal structure. The incident three-beam and outgoing two-beam case is the simplest approximation taking into account elastic scattering both before and after an inelastic scattering event. Adding more beams to the calculation can, of course, improve the results somewhat. However, the effect was found to be very small and well within typical experimental uncertainties [11], owing primarily to the $1/q^2 q'^2$ term in Equation (4) (any additional beams would give rise to much longer q vectors). The crystal structure was assumed to be centro-symmetric, resulting in $U_G = U_{-G}$. While this limits the applicability of the formula to relatively simple crystals, very complex, non-symmetric crystals will likely violate some of the other assumptions as well. In addition, the constraints implied by centro-symmetry are necessary in the first place to arrive at a reasonably simple final formula.

Assumption two requires the sample's surface to be essentially perpendicular to the beam direction. This requirement is necessary to avoid complex phase factors down the line. A small tilt of up to a few degrees is not expected to cause any major issues, and larger tilts of $\gtrsim 45$ °C are not recommended (and often not even possible) in practice anyway.

Assumption three requires the inelastic scattering process to be invariant under rotations around the optical axis by integer multiples of 90°. Strong anisotropy would lead to a distinct directional dependence of the MDFF [48,59,60], thereby making it impossible to reason about the intensities at the various detector positions. In such cases, however, the classical EMCD setup would fail to properly measure the magnetic properties anyway. In addition, assumption three states $q_e \ll |G|$, which implies $\Im[S(q_1, q_2, E)] \ll \Re[S(q_1, q_2, E)]$ in dipole approximation [11,61]. This is fulfilled reasonably well for typical EMCD experiments (for example, for Fe (200), $|G| \approx 7 \, \text{nm}^{-1}$; for the Fe L-edge, $q_e \approx 0.8 \, \text{nm}^{-1}$ at 200 keV and $q_e \approx 1.5 \, \text{nm}^{-1}$ at 40 keV).

Assumption four requires the investigated atoms to be distributed homogeneously and fulfill the condition $G \cdot x = 2m\pi$. The homogeneity requirement excludes involved situations such as multi-layer systems and ultimately allows to replace the sum over all atoms by an integral over the sample thickness. In practice, homogeneity is facilitated by tilting into a systematic row condition and probing a large area of the sample, as a large probed volume and a (small) tilt mean that some atoms can be found in each of the investigated lattice planes at any depth z.

The condition $G \cdot x = 2m\pi \, \forall x$ is perhaps the most severe limitation as it implies that all atoms fall exactly onto one of the probed set of lattice planes. This excludes, e.g., $G = (100)$ for Fe (which is forbidden anyway), or $G = (100)$ for Co, as for these, only some (for Fe) or none (for Co) of the atoms fulfill the condition. The reason for requiring $G \cdot x = 2m\pi$ is that it implies that phase factors of the form $\exp(iG \cdot x)$ are all 1. If that is not the case, different phases have to be applied to different components, thereby reducing the EMCD effect [41]. Hence, choosing a G vector not fulfilling the condition is unfavorable anyway.

As can be seen from Figure 1, Equation (7) reproduces sophisticated numerical simulations quite well for reasonably simple samples despite all approximations. The strongest deviations can be found for small t, as can be expected. For larger sample thicknesses and, consequently, many atoms, small differences that might arise for individual atoms tend to average out.

4. Materials and Methods

The numerical simulations were performed using the "bw" code [11], a software package for calculating EELS data based on Bloch waves and the MDFF. The crystal structure data for magnetite was taken from [62], all other crystallographic data was taken from the EMS program (version 4.5430U2017) [63].

The wedge-shaped magnetite sample was prepared by a FEI Quanta 200 3D DBFIB (FEI Company, Hillsboro, OR, USA) from a high-quality, natural single crystal purchased from SurfaceNet GmbH (Rheine, Germany) [64] and subsequently thinned and cleaned using a Technoorg Linda Gentlemill.

The EMCD measurements were performed on a FEI Tecnai T20 (FEI Company, Hillsboro, OR, USA) equipped with a LaB$_6$ gun and a Gatan GIF 2001 spectrometer (Gatan Inc., Pleasanton, CA, USA). The system has an energy resolution (full width at half maximum) of 1.1 eV at 200 kV which improves down to 0.3 eV at 20 kV [65]. First, a suitable sample position with a sample thickness around 40 nm and an easily recognizable, distinctly-shaped feature nearby was found and the sample was oriented in systematic row condition including the (400) diffraction spot (see Figure 5). At each high tension, the instrument was carefully aligned, the sample position was readjusted, the EMCD experiment was performed, and a thickness measurement was taken. Both the convergence and the collection semi-angle were approximately 3 mrad [58].

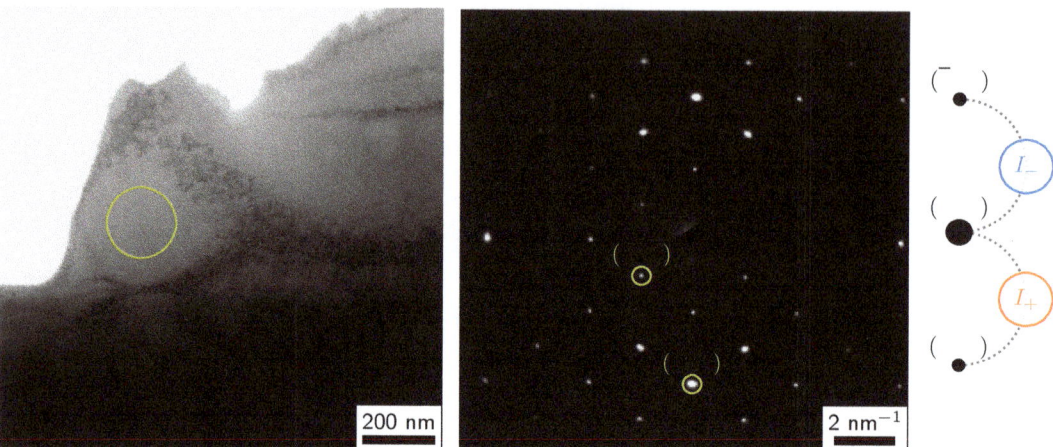

Figure 5. TEM bright-field overview image (**left**), corresponding diffraction pattern in $(0\,1\,\overline{1})$ zone axis (**middle**) and schematic of the EMCD measurement positions in systematic row condition (**right**). The sample position used for the EMCD experiments is marked by a yellow circle in the bright-field image, the positions for I_+ and I_- are marked by the orange and blue circles. Both the image and the diffraction pattern were recorded at 200 kV. Note that the weak, kinematically forbidden $(2\,0\,0)$ reflections can be attributed to double diffraction [36] in the thicker part of the sample visible at the bottom of the bright-field image; they are negligible in the thin part of the sample used for the EMCD measurements.

For data analysis, all spectra were background-subtracted using a pre-edge power-law fit and normalized in the post-edge region. The EMCD effect was calculated based on the L_3-edge maxima according to the formula [9,58]

$$\eta = \frac{I_+ - I_-}{\frac{I_+ + I_-}{2}}. \tag{9}$$

The errors were estimated as described in [57,58].

5. Conclusions

In this work, we have derived an analytical formula for predicting the EMCD effect, taking into account elastic scattering both before and after inelastic scattering events. This formula not only helps elucidate the physics underlying EMCD, it also allows to directly predict the influence of various parameters on the EMCD effect. In particular, we have focused on the acceleration voltage V and on the thickness t. We showed that the periodicity of the EMCD effect scales with $\sqrt{V}$, while its total intensity decreases as $1/t$. In addition, we have performed experiments at different acceleration voltages to corroborate these predictions. Our results will not only help to optimize the EMCD effect for a given sample thickness by tuning the high tension accordingly, it will also pave the way for magnetization-dependent measurements by employing different magnetic fields in the objective lens at different acceleration voltages.

Author Contributions: Conceptualization, S.L., P.S.; methodology, S.L., M.S.-P., W.H., P.S.; software, S.L.; formal analysis, S.L.; investigation, S.L., M.S.-P.; resources, A.S.-T., W.H.; data curation, S.L.; writing—original draft preparation, S.L., P.S.; writing—review and editing, M.S.-P., A.S.-T., W.H.;

visualization, S.L.; supervision, P.S.; project administration, S.L., P.S.; funding acquisition, S.L., P.S. All authors have read and agreed to the published version of the manuscript.

Funding: This research was funded by the Austrian Science Fund (FWF) under grant numbers I4309-N36 and P29687-N36.

Institutional Review Board Statement: Not applicable.

Informed Consent Statement: Not applicable.

Data Availability Statement: Data is contained within the article.

Conflicts of Interest: The authors declare no conflict of interest. The funders had no role in the design of the study; in the collection, analyses, or interpretation of data; in the writing of the manuscript, or in the decision to publish the results.

Abbreviations

The following abbreviations are used in this manuscript:

DDSCS	Double-differential scattering cross-Section
EMCD	Energy-loss magnetic chiral dichroism
EELS	Electron energy-loss spectrometry
MDFF	Mixed dynamic form factor
TEM	Transmission electron microscopy
XAS	X-ray absorption spectroscopy
XMCD	X-ray magnetic circular dichroism

Appendix A. Derivation of the EMCD Effect

In the following, we will extensively use the abbreviations

$$\alpha = \frac{U_G}{2\chi \cdot n} \qquad \tilde{\alpha} = \frac{U_G}{2\tilde{\chi} \cdot \tilde{n}} \tag{A1}$$

$$V = \frac{U_{2G} - |G|^2}{2U_G} \tag{A2}$$

$$W = \frac{\sqrt{(|G|^2 - U_{2G})^2 + 8U_G^2}}{2U_G} = \sqrt{V^2 + 2}, \tag{A3}$$

where the U_g are the Fourier coefficients of the crystal potential $V(r) = \frac{\hbar^2}{2me}\sum_g U_g e^{2\pi i g \cdot r}$ with Planck's constant h, electron mass m and elementary charge e. We note in passing that in the present case, $U_G = U_G^* = U_{-G}$.

With these abbreviations and the assumptions mentioned above, the Bloch wave parameters can be calculated analytically and take the form

$$\begin{array}{lll}
\gamma_1 = \alpha(V+W) & \gamma_2 = \alpha(V-W) & \gamma_3 = -\alpha \cdot \frac{|G|^2 + U_{2G}}{U_G} \\
C_{1,-G} = \frac{1}{\sqrt{|V-W|^2+2}} & C_{2,-G} = \frac{1}{\sqrt{|V+W|^2+2}} & C_{3,-G} = -\frac{1}{\sqrt{2}} \\
C_{1,0} = -\frac{V-W}{\sqrt{|V-W|^2+2}} & C_{2,0} = -\frac{V+W}{\sqrt{|V+W|^2+2}} & C_{3,0} = 0 \\
C_{1,G} = \frac{1}{\sqrt{|V-W|^2+2}} & C_{2,G} = \frac{1}{\sqrt{|V+W|^2+2}} & C_{3,G} = \frac{1}{\sqrt{2}}
\end{array} \tag{A4}$$

for $|\psi_{in}\rangle$ and

$$\begin{array}{ll}
\tilde{\gamma}_1 = \tilde{\alpha} & \tilde{\gamma}_2 = -\tilde{\alpha} \\
\tilde{C}_{1,0} = \frac{1}{\sqrt{2}} & \tilde{C}_{2,0} = \frac{1}{\sqrt{2}} \\
\tilde{C}_{1,G} = \frac{1}{\sqrt{2}} & \tilde{C}_{2,G} = -\frac{1}{\sqrt{2}}
\end{array} \tag{A5}$$

for $|\psi_{\text{out}}\rangle$.

Inserting Equations (5) and (6) into Equation (4), evaluating the integrals, collecting all terms with the same Bloch wave index, and neglecting the weak dependence of $S(q, q', E)/(q^2 q'^2)$ on j, j', l, l' [8,41,55] yields

$$p = \sum_x \sum_{g,g',h,h'} D_g D_{g'}^* \tilde{D}_h^* \tilde{D}_{h'} e^{i(g-g'-h+h') \cdot x} \frac{S(q, q', E)}{q^2 q'^2} \quad (A6)$$

with

$$D_g = \sum_j C_{j,0}^* C_{j,g} e^{i \gamma_j n \cdot x} \qquad \tilde{D}_g = \sum_l \tilde{C}_{l,0}^* e^{-i \tilde{\gamma}_l t} \tilde{C}_{l,h} e^{i \tilde{\gamma}_l \tilde{n} \cdot x} \quad (A7)$$

and

$$q = \Delta \chi + g - h \qquad q' = \Delta \chi + g' - h' \qquad \Delta \chi = \chi - \tilde{\chi}. \quad (A8)$$

Direct summation results in

$$D_{-G} = D_G = \frac{i}{W} e^{i\alpha V n \cdot x} \sin(\alpha W n \cdot x)$$

$$D_0 = e^{i\alpha V n \cdot x} \left[\cos(\alpha W n \cdot x) - \frac{iV}{W} \sin(\alpha W n \cdot x) \right] \quad (A9)$$

$$\tilde{D}_0 = \cos(\tilde{\alpha}(\tilde{n} \cdot x - t))$$

$$\tilde{D}_G = i \sin(\tilde{\alpha}(\tilde{n} \cdot x - t)).$$

Performing the complete sums over g, g', h, h' in Equation (A6) produces very many terms, some of which are very small. This can be understood from the fact that $\Delta \chi \cdot G = \pm G/2$ in the chosen setup. Therefore, $\Delta \chi$ and $\Delta \chi - G$ have the same magnitude, whereas $\Delta \chi + G$ and $\Delta \chi - 2G$ are significantly larger. Owing to the $1/q^2 q'^2$ term, large q are strongly suppressed. Hence, only the combinations $g - h = 0$ and $g - h = -G$ are retained (the same applies to the primed versions as well). Hence, we end up with two distinct q vectors, namely

$$q_1 = \Delta \chi \quad \text{and} \quad q_2 = \Delta \chi - G. \quad (A10)$$

Note that, due to the symmetry of the setup $q_1 = |q_1| = |q_2| = q_2$.
Using $S(q, q', E) = S(q', q, E)^*$ [45], Equation (A6) now takes the form

$$p = \frac{1}{q_1^4} \sum_x [|D_0 \tilde{D}_0^* + D_G \tilde{D}_G^*|^2 S(q_1, q_1, E) +$$

$$|D_{-G} \tilde{D}_0^* + D_0 \tilde{D}_G^*|^2 S(q_2, q_2, E) +$$

$$2\Re \left[(D_0 \tilde{D}_0^* + D_G \tilde{D}_G^*)(D_{-G}^* \tilde{D}_0 + D_0^* \tilde{D}_G) e^{iG \cdot x} S(q_1, q_2, E) \right]] \quad (A11)$$

$$= \frac{1}{q_1^4} [A_{11} S(q_1, q_1, E) + A_{22} S(q_2, q_2, E) + 2\Re[A_{12} S(q_1, q_2, E)]]$$

$$= \frac{1}{q_1^4} [(A_{11} + A_{22}) S(q_1, q_1, E) + 2\Re[A_{12} S(q_1, q_2, E)]].$$

In the last line, the four-fold rotational symmetry was used, i.e., $S(q_1, q_1, E) = S(q_2, q_2, E)$ since $q_2 = \hat{C}_4[q_1]$ with $\hat{C}_4$ as the operator performing a 90° rotation around the optical axis.

To calculate the probability for a "click" in the detector at the second EMCD position, we have to replace $q_1 \mapsto \hat{C}_4^3[q_1] = \hat{C}_4^2[q_2]$ and $q_2 \mapsto \hat{C}_4[q_2] = \hat{C}_4^2[q_1]$. Owing to the assumed rotational symmetry of the MDFF, this replacement results in $S(\hat{C}_4^2[q_2], \hat{C}_4^2[q_1], E) = S(q_2, q_1, E) = S(q_1, q_2, E)^*$ and hence

$$p' = \frac{1}{q_1^4} [(A_{11} + A_{22}) S(q_1, q_1, E) + 2\Re[A_{12} S(q_1, q_2, E)^*]]. \quad (A12)$$

Thus, the quotient EMCD effect is

$$\eta = 2 \cdot \frac{p - p'}{p + p'} = 2 \cdot \frac{-2\Im[A_{12}]\Im[S(q_1, q_2, E)]}{(A_{11} + A_{22})S(q_1, q_1, E) + 2\Re[A_{12}]\Re[S(q_1, q_2, E)]} \quad (A13)$$

Assuming that the scattering vectors were chosen such that $S(q_1, q_2, E)$ is purely imaginary (technically, in dipole approximation) this occurs slightly inside the Thales circle where $q_y^2 = G^2/4 - q_e^2$; as $q_e \ll G$ in typical EMCD experiments, the real part of $S(q_1, q_2, E)$, which is of the order q_e^2, can be neglected compared to $S(q_1, q_1, E)$, which is of the order of $G^2/2$), this can be simplified further to

$$= -4 \cdot \frac{\Im[A_{12}]}{A_{11} + A_{22}} \cdot \frac{\Im[S(q_1, q_2, E)]}{S(q_1, q_1, E)}. \quad (A14)$$

The coefficients can be calculated directly as

$$A_{11} + A_{22} = \sum_x \left[1 - \frac{1}{W^2} \sin^2(\alpha W n \cdot x)\right]$$

$$\Im[A_{12}] = \sum_x \frac{1}{2}\left[\left(1 - \frac{3}{W^2} \sin^2(\alpha W n \cdot x)\right) \sin(2\tilde{\alpha}(\tilde{n} \cdot x - t)) \right. \quad (A15)$$

$$\left. - \frac{1}{W} \sin(2\alpha W n \cdot x) \cos(2\tilde{\alpha}(\tilde{n} \cdot x - t))\right].$$

with the assumptions 2 and 4, the dot products can be evaluated and the sums can be replaced by integrals over z, yielding

$$A_{11} + A_{22} = t\left(1 - \frac{1}{2W^2}\right) + \frac{\sin(2tW\alpha)}{4W^3\alpha}$$

$$\Im[A_{12}] = \frac{1}{4(W^2\alpha^2 - \tilde{\alpha}^2)} \left[-\left(2\alpha + \frac{3\tilde{\alpha}}{W^2}\right)\sin^2(\alpha W t) \right. \quad (A16)$$

$$\left. + \left(\frac{(3 - 2W^2)\alpha^2}{\tilde{\alpha}} + 2(\alpha + \tilde{\alpha})\right)\sin^2(\tilde{\alpha}t)\right]$$

Hence the full formula for the EMCD effect reads

$$\eta = \frac{4W^3\alpha}{(W^2\alpha^2 - \tilde{\alpha}^2)} \cdot \frac{\left[\left(2\alpha + \frac{3\tilde{\alpha}}{W^2}\right)\sin^2(\alpha W t) - \left(\frac{(3-2W^2)\alpha^2}{\tilde{\alpha}} + 2(\alpha + \tilde{\alpha})\right)\sin^2(\tilde{\alpha}t)\right]}{2W(2W^2 - 1)\alpha t + \sin(2tW\alpha)} \cdot \frac{\Im[S(q_1, q_2, E)]}{S(q_1, q_1, E)}$$

$$= \frac{A \sin^2(\kappa t) - B\sin^2(\kappa' t)}{t + C\sin(2\kappa t)} \cdot \frac{\Im[S(q_1, q_2, E)]}{S(q_1, q_1, E)} \quad (A17)$$

with

$$A = C \cdot \frac{4\kappa\kappa'}{\kappa^2 - \kappa'^2}\left(2W\frac{\kappa}{\kappa'} + 3\right)$$

$$B = C \cdot \frac{4\kappa\kappa'}{\kappa^2 - \kappa'^2}\left(2W\frac{\kappa}{\kappa'} + \frac{3\kappa^2}{\kappa'^2} + 2W^2\left(1 - \frac{\kappa^2}{\kappa'^2}\right)\right)$$

$$C = \frac{1}{2\kappa(2W^2 - 1)} \quad (A18)$$

$$\kappa = \alpha W = \frac{\gamma_1 - \gamma_2}{2}$$

$$\kappa' = \tilde{\alpha} = \frac{\tilde{\gamma}_1 - \tilde{\gamma}_2}{2}.$$

References

1. Botton, G. A new approach to study bonding anisotropy with EELS. *J. Electron Spectrosc. Relat. Phenom.* **2005**, *143*, 129–137. [CrossRef]
2. Schattschneider, P.; Hébert, C.; Franco, H.; Jouffrey, B. Anisotropic relativistic cross sections for inelastic electron scattering, and the magic angle. *Phys. Rev. B* **2005**, *72*, 045142. [CrossRef]
3. Hébert, C.; Schattschneider, P. A proposal for dichroic experiments in the electron microscope. *Ultramicroscopy* **2003**, *96*, 463–468. [CrossRef]
4. Schattschneider, P.; Jouffrey, B. Channeling, localization and the density matrix in inelastic electron scattering. *Ultramicroscopy* **2003**, *96*, 453–462. [CrossRef]
5. Schattschneider, P.; Werner, W. Coherence in electron energy loss spectrometry. *J. Electron Spectrosc. Relat. Phenom.* **2005**, *143*, 81–95. [CrossRef]
6. CHIRALTEM. Available online: https://cordis.europa.eu/project/id/508971 (accessed on 9 of February, 2021).
7. Schattschneider, P.; Rubino, S.; Hebert, C.; Rusz, J.; Kunes, J.; Novák, P.; Carlino, E.; Fabrizioli, M.; Panaccione, G.; Rossi, G. Detection of magnetic circular dichroism using a transmission electron microscope. *Nature* **2006**, *441*, 486–488. [CrossRef]
8. Rusz, J.; Rubino, S.; Schattschneider, P. First-principles theory of chiral dichroism in electron microscopy applied to 3d ferromagnets. *Phys. Rev. B* **2007**, *75*, 214425. [CrossRef]
9. Hébert, C.; Schattschneider, P.; Rubino, S.; Novak, P.; Rusz, J.; Stöger-Pollach, M. Magnetic circular dichroism in electron energy loss spectrometry. *Ultramicroscopy* **2008**, *108*, 277–284. [CrossRef]
10. Verbeeck, J.; Hébert, C.; Rubino, S.; Novák, P.; Rusz, J.; Houdellier, F.; Gatel, C.; Schattschneider, P. Optimal aperture sizes and positions for EMCD experiments. *Ultramicroscopy* **2008**, *108*, 865–872. [CrossRef] [PubMed]
11. Löffler, S.; Schattschneider, P. A software package for the simulation of energy-loss magnetic chiral dichroism. *Ultramicroscopy* **2010**, *110*, 831–835. [CrossRef] [PubMed]
12. Rusz, J. Modified automatic term selection v2: A faster algorithm to calculate inelastic scattering cross-sections. *Ultramicroscopy* **2017**, *177*, 20–25. [CrossRef]
13. Schattschneider, P.; Stöger-Pollach, M.; Rubino, S.; Sperl, M.; Hurm, C.; Zweck, J.; Rusz, J. Detection of magnetic circular dichroism on the two-nanometer scale. *Phys. Rev. B* **2008**, *78*, 104413. [CrossRef]
14. Schneider, S.; Pohl, D.; Löffler, S.; Rusz, J.; Kasinathan, D.; Schattschneider, P.; Schultz, L.; Rellinghaus, B. Magnetic properties of single nanomagnets: Electron energy-loss magnetic chiral dichroism on FePt nanoparticles. *Ultramicroscopy* **2016**, *171*, 186–194. [CrossRef]
15. Verbeeck, J.; Schattschneider, P.; Lazar, S.; Stöger-Pollach, M.; Löffler, S.; Steiger-Thirsfeld, A.; Van Tendeloo, G. Atomic scale electron vortices for nanoresearch. *Appl. Phys. Lett.* **2011**, *99*, 203109. [CrossRef]
16. Schattschneider, P.; Schaffer, B.; Ennen, I.; Verbeeck, J. Mapping spin-polarized transitions with atomic resolution. *Phys. Rev. B* **2012**, *85*, 134422. [CrossRef]
17. Schachinger, T.; Löffler, S.; Steiger-Thirsfeld, A.; Stöger-Pollach, M.; Schneider, S.; Pohl, D.; Rellinghaus, B.; Schattschneider, P. EMCD with an electron vortex filter: Limitations and possibilities. *Ultramicroscopy* **2017**, *179*, 15–23. [CrossRef]
18. Rusz, J.; Muto, S.; Spiegelberg, J.; Adam, R.; Tatsumi, K.; Bürgler, D.E.; Oppeneer, P.M.; Schneider, C.M. Magnetic measurements with atomic-plane resolution. *Nat. Commun.* **2016**, *7*, 12672. [CrossRef]
19. Warot-Fonrose, B.; Houdellier, F.; Hÿtch, M.; Calmels, L.; Serin, V.; Snoeck, E. Mapping inelastic intensities in diffraction patterns of magnetic samples using the energy spectrum imaging technique. *Ultramicroscopy* **2008**, *108*, 393–398. [CrossRef]
20. Salafranca, J.; Gazquez, J.; Pérez, N.; Labarta, A.; Pantelides, S.T.; Pennycook, S.J.; Batlle, X.; Varela, M. Surfactant Organic Molecules Restore Magnetism in Metal-Oxide Nanoparticle Surfaces. *Nano Lett.* **2012**, *12*, 2499–2503. [CrossRef]
21. Thersleff, T.; Rusz, J.; Rubino, S.; Hjörvarsson, B.; Ito, Y.; Zaluzec, N.J.; Leifer, K. Quantitative analysis of magnetic spin and orbital moments from an oxidized iron (1 1 0) surface using electron magnetic circular dichroism. *Sci. Rep.* **2015**, *5*, 13012. [CrossRef]
22. Song, D.; Ma, L.; Zhou, S.; Zhu, J. Oxygen deficiency induced deterioration in microstructure and magnetic properties at $Y_3Fe_5O_{12}$/Pt interface. *Appl. Phys. Lett.* **2015**, *107*, 042401. [CrossRef]
23. Wang, Z.; Tavabi, A.H.; Jin, L.; Rusz, J.; Tyutyunnikov, D.; Jiang, H.; Moritomo, Y.; Mayer, J.; Dunin-Borkowski, R.E.; Yu, R.; et al. Atomic scale imaging of magnetic circular dichroism by achromatic electron microscopy. *Nature Materials* **2018**, *17*, 221–225. [CrossRef]
24. Wang, Z.; Zhong, X.; Yu, R.; Cheng, Z.; Zhu, J. Quantitative experimental determination of site-specific magnetic structures by transmitted electrons. *Nat. Commun.* **2013**, *4*, 1395. [CrossRef]
25. Ennen, I.; Löffler, S.; Kübel, C.; Wang, D.; Auge, A.; Hütten, A.; Schattschneider, P. Site-specific chirality in magnetic transitions. *J. Magn. Magn. Mater.* **2012**, *324*, 2723–2726. [CrossRef]
26. Loukya, B.; Negi, D.S.; Dileep, K.; Pachauri, N.; Gupta, A.; Datta, R. Effect of Bloch wave electron propagation and momentum-resolved signal detection on the quantitative and site-specific electron magnetic chiral dichroism of magnetic spinel oxide thin films. *Phys. Rev. B Condens. Matter Mater. Phys.* **2015**, *91*, 134412. [CrossRef]
27. Wallisch, W.; Stöger-Pollach, M.; Navickas, E. Consequences of the CMR effect on EELS in TEM. *Ultramicroscopy* **2017**, *179*, 84–89. [CrossRef]

28. Wang, Z.C.; Zhong, X.Y.; Jin, L.; Chen, X.F.; Moritomo, Y.; Mayer, J. Effects of dynamic diffraction conditions on magnetic parameter determination in a double perovskite Sr_2FeMoO_6 using electron energy-loss magnetic chiral dichroism. *Ultramicroscopy* **2017**, *176*, 212–217. [CrossRef]
29. Fu, X.; Warot-Fonrose, B.; Arras, R.; Dumesnil, K.; Serin, V. Quantitative moment study and coupling of 4f rare earth and 3d metal by transmitted electrons. *Phys. Rev. B* **2016**, *94*, 140416. [CrossRef]
30. Lin, J.; Zhong, X.Y.; Song, C.; Rusz, J.; Kocevski, V.; Xin, H.L.; Cui, B.; Han, L.L.; Lin, R.Q.; Chen, X.F.; Zhu, J. Detection of magnetic circular dichroism in amorphous materials utilizing a single-crystalline overlayer. *Phys. Rev. Mater.* **2017**, *1*, 071404. [CrossRef]
31. He, M.; He, X.; Lin, L.; Song, B.; Zhang, Z.H. Study on spin polarization of non-magnetic atom in diluted magnetic semiconductor: The case of Al-doped 4H-SiC. *Solid State Commun.* **2014**, *197*, 44–48. [CrossRef]
32. Fu, X.; Warot-Fonrose, B.; Arras, R.; Seine, G.; Demaille, D.; Eddrief, M.; Etgens, V.; Serin, V. In situ observation of ferromagnetic order breaking in MnAs/GaAs(001) and magnetocrystalline anisotropy of α-MnAs by electron magnetic chiral dichroism. *Phys. Rev. B* **2016**, *93*, 104410. [CrossRef]
33. Chen, X.; Higashikozono, S.; Ito, K.; Jin, L.; Ho, P.; Yu, C.; Tai, N..; Mayer, J.; Dunin-Borkowski, R.E.; Suemasu, T.; Zhong, X. Nanoscale measurement of giant saturation magnetization in α''-$Fe_{16}N_2$ by electron energy-loss magnetic chiral dichroism. *Ultramicroscopy* **2019**, *203*, 37–43. [CrossRef] [PubMed]
34. Stöger-Pollach, M.; Treiber, C.D.; Resch, G.P.; Keays, D.A.; Ennen, I. EMCD real space maps of Magnetospirillum magnetotacticum. *Micron* **2011**, *42*, 456–460. [CrossRef] [PubMed]
35. Schattschneider, P., Ed. *Linear and Chiral Dichroism in the Electron Microscope*; Pan Stanford Publishing Pte Ltd.: Singapore, 2011.
36. Williams, D.B.; Carter, C.B. *Transmission Electron Microscopy*; Plenum Press: New York, NY, USA, 1996.
37. Egerton, R.; Li, P.; Malac, M. Radiation damage in the TEM and SEM. *Micron* **2004**, *35*, 399–409. [CrossRef] [PubMed]
38. Egerton, R. Control of radiation damage in the TEM. *Ultramicroscopy* **2013**, *127*, 100–108. [CrossRef] [PubMed]
39. Jiang, N. Electron beam damage in oxides: A review. *Rep. Prog. Phys.* **2015**, *79*, 016501. [CrossRef] [PubMed]
40. Hurm, C. Towards an Unambiguous Electron Magnetic Chiral Dichroism (EMCD) Measurement in a Transmission Electron Microscope (TEM). Ph.D. Thesis, Universität Regensburg, Regensburg, Germany, 2008.
41. Schneider, S.; Negi, D.; Stolt, M.J.; Jin, S.; Spiegelberg, J.; Pohl, D.; Rellinghaus, B.; Goennenwein, S.T.B.; Nielsch, K.; Rusz, J. Simple method for optimization of classical electron magnetic circular dichroism measurements: The role of structure factor and extinction distances. *Phys. Rev. Mater.* **2018**, *2*, 113801. [CrossRef]
42. Rusz, J.; Novák, P.; Rubino, S.; Hébert, C.; Schattschneider, P. Magnetic Circular Dichroism in Electron Microscopy. *Acta Phys. Pol. A* **2008**, *113*, 599–644. [CrossRef]
43. Pogany, A.P.; Turner, P.S. Reciprocity in electron diffraction and microscopy. *Acta Cryst. A* **1968**, *24*, 103–109. [CrossRef]
44. Findlay, S.; Schattschneider, P.; Allen, L. Imaging using inelastically scattered electrons in CTEM and STEM geometry. *Ultramicroscopy* **2007**, *108*, 58–67. [CrossRef]
45. Kohl, H.; Rose, H. Theory of Image Formation by Inelastically Scattered Electrons in the Electron Microscope. *Adv. Electron. Electron Phys.* **1985**, *65*, 173–227. [CrossRef]
46. Schattschneider, P. *Fundamentals of Inelastic Electron Scattering*; Springer Wien: New York, NY, USA, 1986.
47. Nelhiebel, M. Effects of Crystal Orientation and Interferometry in Electron Energy Loss Spectroscopy. Ph.D. Thesis, École Centrale Paris, Châtenay-Malabry, France, 1999.
48. Löffler, S.; Motsch, V.; Schattschneider, P. A pure state decomposition approach of the mixed dynamic form factor for mapping atomic orbitals. *Ultramicroscopy* **2013**, *131*, 39–45. . [CrossRef]
49. Löffler, S. Study of Real Space Wave Functions with Electron Energy Loss Spectrometry. Ph.D. Thesis, TU Wien, Vienna, Austria, 2013.
50. Schattschneider, P.; Nelhiebel, M.; Jouffrey, B. Density matrix of inelastically scattered fast electrons. *Phys. Rev. B* **1999**, *59*, 10959–10969. [CrossRef]
51. Schattschneider, P.; Nelhiebel, M.; Souchay, H.; Jouffrey, B. The physical significance of the mixed dynamic form factor. *Micron* **2000**, *31*, 333–345. [CrossRef]
52. Schattschneider, P.; Ennen, I.; Löffler, S.; Stöger-Pollach, M.; Verbeeck, J. Circular dichroism in the electron microscope: Progress and applications (invited). *J. Appl. Phys.* **2010**, *107*, 09D311. [CrossRef]
53. Stadelmann, P. *Dynamical Theory of Elastic Electron Diffraction at Small Angles*; Technical Report; École Polytechnique Fédérale de Lausanne: Lausanne, Switzerland, 2003.
54. Metherell, A. J., F. Diffraction of Electrons by Perfect Crystals. In *Electron Microscopy in Materials Science*; Valdrè, U.; Ruedl, E., Eds.; Commission of the European Communities: Brussels, Belgium, 1975; Volume 2, pp. 397–552.
55. Schattschneider, P.; Jouffrey, B.; Nelhiebel, M. Dynamical diffraction in electron-energy-loss spectrometry: The independent Bloch-wave model. *Phys. Rev. B* **1996**, *54*, 3861–3868. [CrossRef] [PubMed]
56. Hetaba, W. The Theory and Application of Inelastic Coherence in the Electron Microscope. Ph.D. Thesis, TU Wien, Vienna, Austria 2015.
57. Egerton, R.F. Electron energy-loss spectroscopy in the TEM. *Rep. Prog. Phys.* **2009**, *72*, 016502. [CrossRef]
58. Löffler, S.; Hetaba, W. Convergent-beam EMCD: Benefits, pitfalls and applications. *Microscopy* **2018**, *67*, i60–i71. [CrossRef]

59. Löffler, S.; Bugnet, M.; Gauquelin, N.; Lazar, S.; Assmann, E.; Held, K.; Botton, G.A.; Schattschneider, P. Real-space mapping of electronic orbitals. *Ultramicroscopy* **2017**, *177*, 26–29. [CrossRef]
60. Löffler, S.; Hambach, R.; Kaiser, U.; Schattschneider, P. Symmetry-constraints for mapping electronic states with EELS. **2021**, in preparation.
61. Schattschneider, P.; Hébert, C.; Rubino, S.; Stöger-Pollach, M.; Rusz, J.; Novák, P. Magnetic circular dichroism in EELS: Towards 10 nm resolution. *Ultramicroscopy* **2008**, *108*, 433–438. [CrossRef] [PubMed]
62. Fleet, M.E. The structure of magnetite. *Acta Crystallogr. Sect. B Struct. Crystallogr. Cryst. Chem.* **1981**, *37*, 917–920. [CrossRef]
63. Stadelmann, P. EMS—A software package for electron diffraction analysis and HREM image simulation in materials science. *Ultramicroscopy* **1987**, *21*, 131–145. [CrossRef]
64. SurfaceNet GmbH. Oskar-Schindler-Ring 7, 48432 Rheine, Germany. Available online: https://www.surfacenet.de/ (accessed on 9 of February, 2021).
65. Stöger-Pollach, M. Low voltage TEM: Influences on electron energy loss spectrometry experiments. *Micron* **2010**, *41*, 577–584. [CrossRef]

MDPI
St. Alban-Anlage 66
4052 Basel
Switzerland
Tel. +41 61 683 77 34
Fax +41 61 302 89 18
www.mdpi.com

Materials Editorial Office
E-mail: materials@mdpi.com
www.mdpi.com/journal/materials

www.ingramcontent.com/pod-product-compliance
Lightning Source LLC
LaVergne TN
LVHW070557100526
838202LV00012B/487